T0325712

Food Safety and Quality Systems in Developing Countries

Food Safety and Quality Systems in Developing Countries

Volume One: Export Challenges and Implementation Strategies

Edited by

André Gordon

Technological Solutions Limited, Kingston, Jamaica

AMSTERDAM • BOSTON • HEIDELBERG • LONDON
NEW YORK • OXFORD • PARIS • SAN DIEGO
SAN FRANCISCO • SINGAPORE • SYDNEY • TOKYO

Academic Press is an imprint of Elsevier

Academic Press is an imprint of Elsevier
125, London Wall, EC2Y 5AS, UK
525 B Street, Suite 1800, San Diego, CA 92101-4495, USA
225 Wyman Street, Waltham, MA 02451, USA
The Boulevard, Langford Lane, Kidlington, Oxford OX5 1GB, UK

Notices
Knowledge and best practice in this field are constantly changing. As new research and experience broaden our understanding, changes in research methods, professional practices, or medical treatment may become necessary.

Practitioners and researchers must always rely on their own experience and knowledge in evaluating and using any information, methods, compounds, or experiments described herein. In using such information or methods they should be mindful of their own safety and the safety of others, including parties for whom they have a professional responsibility.

To the fullest extent of the law, neither the Publisher nor the authors, contributors, or editors, assume any liability for any injury and/or damage to persons or property as a matter of products liability, negligence or otherwise, or from any use or operation of any methods, products, instructions, or ideas contained in the material herein.

British Library Cataloguing-in-Publication Data
A catalogue record for this book is available from the British Library

Library of Congress Cataloging-in-Publication Data
A catalog record for this book is available from the Library of Congress

ISBN: 978-0-12-801227-7

For information on all Academic Press publications
visit our website at http://store.elsevier.com/

Publisher: Shirley Decker-Lucke
Acquisition Editor: Patricia Osborn
Editorial Project Manager: Jaclyn Truesdell
Production Project Manager: Julia Haynes
Designer: Matt Limbert

Typeset by Thomson Digital
Printed and bound in the USA

CONTENTS

LIST OF CONTRIBUTORS

Orane Blake
A. T. Kearney Inc, Dallas, Texas, USA

André Gordon
Technological Solutions Limited, Kingston, Jamaica

Jose Jackson-Malete
Botswana Institute for Technology Research and Innovation (BITRI), Gaborone, Botswana

James Kerr
Retired from the Bureau of Standards Jamaica, Kingston, Jamaica

Joyce Saltsman
Retired from the Center for Food Safety and Applied Nutrition (CFSAN), US Food and Drug Administration, Baltimore, Maryland, USA

George Ware
Retired from the Southeast Regional Laboratory (SRL), US Food and Drug Administration, Atlanta, USA

PREFACE

The importance of developing countries as providers of an increasing percentage of the food being consumed globally is receiving growing recognition. This is because these countries remain an undeniably important source of key food items for developed-country consumers, including increasingly sophisticated food product offerings. Consequently, the status of the food safety and quality systems in these countries is no longer a matter of local interest only; a food safety challenge in Asia or Africa can have repercussions as far away as Paris, Frankfurt, Los Angeles, Rio de Janeiro, or Vancouver. Recognizing this, we felt it was timely to provide a different perspective on food safety and quality systems with a focus on exports from developing countries, and with important lessons for many stakeholders in the global food supply chain, regardless of origin.

The approach we have taken is to focus on practical trade and market access-related considerations and to underpin the analysis and proposed solutions by sound science. This we have presented using a case-study approach, providing historically accurate details of how very difficult export challenges with a fruit have been successfully addressed using food-science-based approaches, including research, where appropriate. This is augmented by several other examples of specific food safety concerns for a range of nontraditional fruit and vegetable products of increasing economic importance that are being imported into developed-country markets. While we have largely focused on the United States in this volume, we do also discuss general principles and look at examples involving the United Kingdom and Canada.

We start the book by examining the food science and trade-related technical considerations in exporting traditional fruits and vegetables to the United States and end it with a comprehensive look at the US Food Safety Modernization Act, 2011. In between this, we develop the concept of applying food science and technology – including risk analysis, research and, where appropriate, the adaptation and implementation of hazard analysis critical control points (HACCP)-based food safety systems – to addressing export challenges. We present detailed descriptions of very specific market access challenges of a technical or scientific nature for selected products and examine the use of science and research to mitigate these. Also, for the first time, we provide a detailed scientific treatment for the production, commercial handling and processing, culinary and consumption information, biochemistry, and export of the fruit *Blighia sapida* from multiple countries. The use of this information in solving multiple market access challenges in three markets is one of the highlights of this book.

This volume is targeted at practitioners of food safety and quality in regulatory bodies, food handling and exporting firms, other exporters, policy makers in governments and their international development agency partners, market access specialists, and food science and technology students and academia. It is expected that people within the target audience in both developing and developed countries will gain valuable insights into the nature of export challenges and potential implementation strategies. The editor has been at pains to ensure that, while dealing with foods from developing countries and the associated challenges, the information presented gives a holistic picture from both the importing and exporting country/firm perspectives. This is reflected in the composition of our authors, who are associated with firms and organizations in both developed and developing countries and are from the private sector, academia, and regulatory institutions. Their backgrounds include strong analytical, applied research, regulatory, and market access and trade expertise, a unique mix for this unique treatment of the subject at hand. As a group, our authors have over 160 years of experience in research, academia, industry, export, trade, and regulatory affairs in North America, Africa, the European Union, and Latin America and the Caribbean. Collectively, they have dealt with fruit and vegetable products from all parts of the world.

This book has arisen from a fortuitous meeting with Ms. Patricia Osborn, Senior Acquisitions Editor, Food Science & Technology, at Elsevier at the International Association for Food Protection (IAFP) Meeting in 2013, where I was an invited speaker. We hope that readers will find in this volume a treasure trove of information, approaches to assessing export market access challenges, and effective collaborative ways to meet them. We also hope that scientists and trade practitioners will get a clear understanding of how food science and technology are playing, and will continue to play, a role in expanding food choices globally while assisting emerging markets to transition into prosperity.

André Gordon
February 20, 2015

ACKNOWLEDGMENTS

This book has been created through the collaboration of many people who have worked together to make it happen. It would perhaps never have been written at all if Ms. Patricia Osborn, Senior Acquisitions Editor, Food Science & Technology, at Elsevier Books Division had not had the confidence to give me an opportunity to create what we hope is an important contribution to the literature on food safety and quality systems. It certainly would not have been completed without the support, guidance, and encouragement first of Patricia, then Carrie Bolger, Editorial Project Manager, Life Sciences, and finally of Marisa LaFleur, Editorial Project Manager, Elsevier. Marisa added the element of focused time management to ensure that all of the contributors and, critically, the editor, stayed on track.

This volume has been blessed to have the contributions of some very special people, my coauthors on various chapters, who bring to it their significant practical and academic knowledge, as well as their unique perspectives. I have worked with each of them individually and collectively on a range of important projects, including research, systems implementation, and training, and particularly on aspects of the transformational work used as a case study in this book. I am deeply honored that they accepted my invitation to participate in this project. I am highly appreciative of the collaboration and contributions of Drs. Jose Jackson, Orane Blake, and Joyce Saltsman, and Messrs. James Kerr and George Ware, who have helped to make this work the unique and informative volume that it is.

Thanks to the people that proofread the book to ensure the identification of areas needing attention. Thanks also to my support team at Technological Solutions Limited (TSL), my company, who took some of the pressure off me, allowing me the time to complete this volume. Finally, I must express my deep gratitude to my immediate and extended family who supported and encouraged me throughout this process, particularly my mother Cynthia Gordon, and my wife Dianne Gordon, both of whom reviewed aspects of the manuscript and gave invaluable feedback. This work would not have been possible without the support provided by all concerned.

Thank you all.

André Gordon, PhD, CFS
February 18, 2015

CHAPTER 1

Exporting Traditional Fruits and Vegetables to the United States: Trade, Food Science, and Sanitary and Phytosanitary/Technical Barriers to Trade Considerations

André Gordon

Technological Solutions Limited, Kingston, Jamaica

ABSTRACT

Despite the challenges that have to be overcome by exporters from developing countries seeking to send traditional foods into developed-country markets, current world trends provide significant opportunities for growth in this area. This chapter shows that the demand in developed countries for traditional fruits and vegetables from developing countries is very favorable and is expected to continue to be strong, including in the United States, the market of major focus. This chapter examines the growing trade in fruits such as soursop (*Annona muricata*), scotch bonnet peppers (*Capsicum frutescens*), and June plum (*Spondias dulcis*) and vegetables such as *Amaranthus* sp. and the reasons for this growth. The increasing demand has been very positive for the growth of the food industry in developing countries and this has widened and deepened the

Food Safety and Quality Systems in Developing Countries. http://dx.doi.org/10.1016/B978-0-12-801227-7.00001-9

options for food industry professionals who have, in turn, further accelerated the development of the sector. The importance of sanitary and phytosanitary (SPS) measures and technical barriers to trade (TBT) to exporters and the need to comply with them is discussed, as is the role of industry specialists in enhancing the ability of emerging economies to actively and successfully participate in trade with their developed-country partners.

Keywords: developing countries; SPS and TBT; fruits and vegetables; *Spondias dulcis*; *Amaranthus* sp.; *Annona muricata*

1.1 INTRODUCTION

Globally, the demand for fruits and vegetables has increased significantly, with developed-country staples such as grapes, melons, citrus fruits, apples and other tree fruits, and fresh tomatoes, peppers, potatoes, onions, and cucumbers being among the many items now being routinely imported, a growing percentage of these from developing countries. This greater focus on exports by developing countries, the opening up of developed-country markets, and the better positioning of many developing countries to supply the increasingly diverse dietary and culinary needs of their more developed trading partners have become the driver for economic growth and development in many emerging economies globally. This has also provided opportunities for many in the areas of agriculture, exports, and trade, as well as for a wide range of food-industry professionals in both developed and developing countries to work to further enrich the diet of the more developed countries while contributing toward enhanced diversity and security in the global food supply. This trade has also opened the door for the export of what are now being called non-traditional agricultural exports (NTAEs) from emerging economies to developed countries (Hallam et al., 2004), which has seen an acceleration of what was an already robust rate of diversification of the trade in fruits and vegetables. Products such as dried and canned mushrooms, cassava (*Manihot esculenta*), eggplant (aubergine; *Solanum melongena*), bok choy (*Brassica rapa* subspecies *chinensis*), sorrel (*Rumex acetosa*), callaloo (*Amaranthus dubius*), and mangoes (*Mangifera indica*), and even relatively unknown fruits such as mammee apple (*Mammea americana*) and ackee (*Blighia sapida*), have become a part of this trade.

The countries that have been leading this growth in the global supply of selected NTAEs from the early 2000s (Hallam et al., 2004) include Mexico (peppers, eggplant, onions and shallots, papaya and mangoes), the Philippines (mangoes, processed fruits), Costa Rica (pineapples), China (mushrooms, peppers, fruit preparations, and medicinal plants), and Thailand (fresh fruit, canned pineapples, ginger, and dried fruit). Newer, increasingly dominant exporters like Brazil, Chile, and India, as well as some African countries such as Côte d'Ivoire (pineapples), Kenya (green beans and peas), and South Africa (processed fruits) have emerged more recently (Diop and Jaffee, 2005). The continued expansion of this trend will depend on the ability of the agro-food sector in emerging economies to develop and nurture a mix of skills and competencies to harness the approaches, technologies, and systems that will be required to meet the ever-changing demands of these sophisticated markets and their consumers. In this regard, the role of food and agricultural sciences and the professionals required to apply these technologies will be critical.

A key consideration in export trade is potential sanitary and phytosanitary (SPS)-related obstacles, as well as technical barriers to trade (TBT) that can impede the ability of developing-country exporters to access developed-country markets (Gordon, 2003; Mendonca and Gordon, 2004). This chapter, therefore, will not only examine the nature of the opportunity for trade in fruits and vegetables, but also introduce the issue of approaches to providing solutions to market-access challenges. The US market, which is still the largest market in the world for developing-country food exports, will be used as an example.

1.2 IMPORTS OF FRUITS AND VEGETABLES INTO THE UNITED STATES: COMPOSITION AND SELECTED TRENDS

The nature of the US population and their demands for fruits and vegetables are constantly changing. Consumers in the United States are no longer willing to accept a limitation in their choice of fruits and vegetables to what was traditional 10 years ago. Exposure to a wider variety of cuisines, increasing travel, and greater diversification in the ethnic makeup of most major US cities will ensure that the changes in dietary patterns and the movement toward the importation and wide consumption of what are regarded as "nontraditional" fruits and vegetables will continue. Another factor influencing this change is the large and growing Hispanic population in many urban areas, as well as continuing migration northward from Latin America and the Caribbean. The United States has gone from being a net exporter of fresh and processed fruits and vegetables in the early 1970s to being a net importer of fruits and vegetables today. The total value of US imports since the 1990s has more than tripled, such that in 2011, US imports exceeded US$18 billion (Johnson, 2014). Increases were greatest for tropical fruits, while imports of fresh vegetables also increased across most categories. The top 10 countries supplying the United States with fruits and vegetables (by share of total import value in 2011) are Mexico (36%); Canada (13%); Chile (8%); China (8%); Costa Rica (5%); Guatemala, Ecuador, and Peru (each about 3–4%); and Argentina and Thailand (2% each). Other leading import suppliers were Brazil, Spain, Honduras, and Colombia (for a combined total of 7%). All other exporting countries supplied a combined 9% of US imports.

Between 1980 and 2010, per capita consumption of all fresh and processed fruits and vegetables increased. While fruits such as peppers, bananas, other tropical fruits, melons, citrus fruits, tree fruits, fruit juices, and various fresh and processed products are among the fruits being imported, NTAEs such as mangoes and other fruits and vegetables are also making significant inroads into the US market (CentralAmerica Data.com, 2012a). This has led to imported fresh and processed fruits and fruit-based products becoming dominant in the supply chain in the United States (Figure 1.1). The expansion in consumer choice because of the experiential and demographic factors mentioned above has contributed to overall higher demand for fruits and vegetables.

As consumer demand for fruits and vegetables has grown, the United States has become a growing market for off-season fruit and vegetable imports. The proximity of major regional exporting countries to the United States and opportunities for counter-seasonal supplies, driven in part by this increased domestic and year-round demand for fruits and vegetables, and diversification in market sourcing by US companies are among the drivers for the increased US importation of fruits and vegetables. Also, improvements in transportation and refrigeration have made it easier to ship fresh horticultural products. Counter-seasonal US imports of fruits and vegetables are supplied mainly by Chile, Argentina,

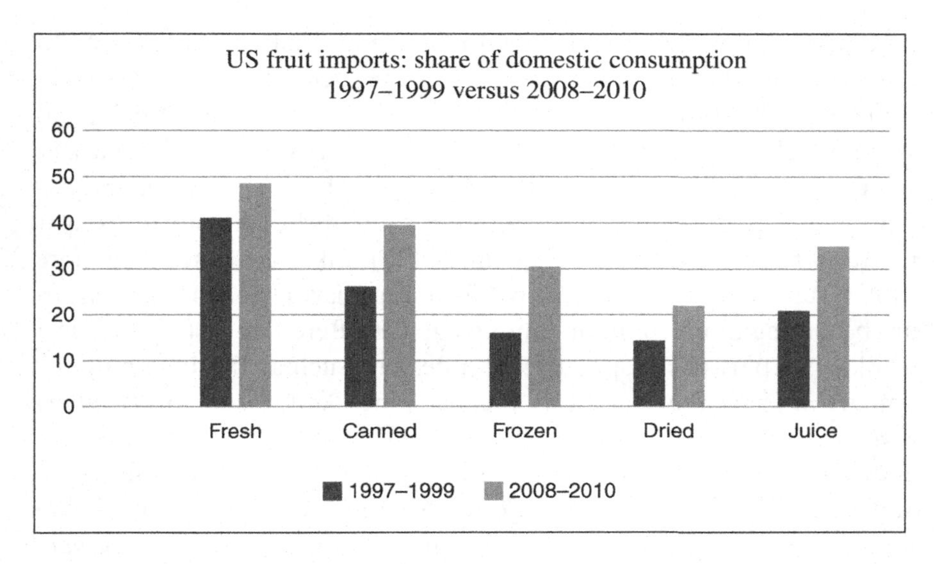

Fig. 1.1. Increasing share of US domestic consumption taken by imported fruits. (Source: USDA Economic Research Service.)

Australia, and South Africa, but also to some extent Mexico and some Central American and Caribbean countries. Counter-seasonal imports from these countries are said to complement US production of fresh grapes, citrus fruits, tree fruits, and berries. However, technological and production improvements are further influencing this trend.

1.3 NONTRADITIONAL TROPICAL FRUIT IMPORTS TO THE UNITED STATES: SPECIFIC EXAMPLES

The tropical fruit category is rapidly becoming pivotal to success in the food service chain in the United States, with suppliers seeing tropical products, including mangoes, becoming increasingly popular and in demand in the food service and retail sectors (CentralAmericaData.com, 2012b). One of the reasons for this is that chefs like tropical fruits because they afford them the versatility to change the look and nature of their meal presentations and offer them a high level of flexibility. A prime example is mango, which has an excellent flavor and color and good nutritional content, and is available throughout the year. Mangoes are starting to compete with other more traditional products, such as strawberries and pineapples, in this market. Mango imports rose from US$226 million in 2007 to US$343 million in 2011, just behind pineapples (approximately US$500 million) and citrus fruits (US$515 million), representing an increase of 13%. Mexico has become the main supplier of the fruit, followed by Peru, Brazil, Ecuador, Guatemala, and Haiti. Other tropical fruits are also finding their way into the market, such as carambola (star fruit), with its unique shape and flavor, and papaya, owing to the ease of incorporating it into dishes directly or using it in sauces (CentralAmericaData.com, 2012b). Likewise, other even more exotic fruits that provide exceptionally unique flavors and characteristics are becoming more common among choices offered by restaurants as a drink or as part of an appetizer or a dessert, such as the June plum or ambarella (*Spondias dulcis*), as well as some like the soursop (*Annona muricata*) (Figure 1.2) that have been traditionally regarded as having additional health benefits. Soursop is also known as *graviola* in Brazil and *guanábana* in Central and South America. The June plum is also known as golden apple (in Barbados and the eastern Caribbean), *pom cythere* (in Trinidad and Tobago), *buah longlong* (in China), and *kedondong* (in Indonesia).

The soursop (Figure 1.2a) has a flavor that has been described as a combination of strawberry and pineapple with a hint of citrus, and is traditionally considered to be among the premier medicinal plants in many Latin American and Caribbean countries. The leaves, fruit, seeds, and stem are used for medicinal purposes and the fruit has an excellent nutritional profile (Figures 1.3a, b and 1.4), the juice being thought to have significant health benefits. The fruit is thought to contain many chemicals that may be active against cancer, as well as against disease-causing agents such as bacteria, viruses, and parasites, although this has not been substantiated (WebMD, 2015). Interestingly, soursop has also been regarded by some sources as potentially unsafe as it might contain an alkaloid-based neurotoxin, annonacin, that is concentrated in the seeds and could cause neuronal dysfunction and degeneration leading to symptoms of Parkinson's disease. This has, however, never been proven, nor has there been a confirmed case of illness that has been directly linked to soursop. Its composition and the potential health risk associated with it make it a very interesting fruit for future study.

The June plum, also known as ambarella (and Malay apple in the Far East), is found throughout Latin America and the Caribbean as well as in the Asian tropics, and is oval in shape, green to yellow-gold in

*Fig. 1.2. Two traditional developing-country fruits now being exported: (a) soursop (*A. muricata*) and (b) June plum (ambarella;* S. dulcis*).*

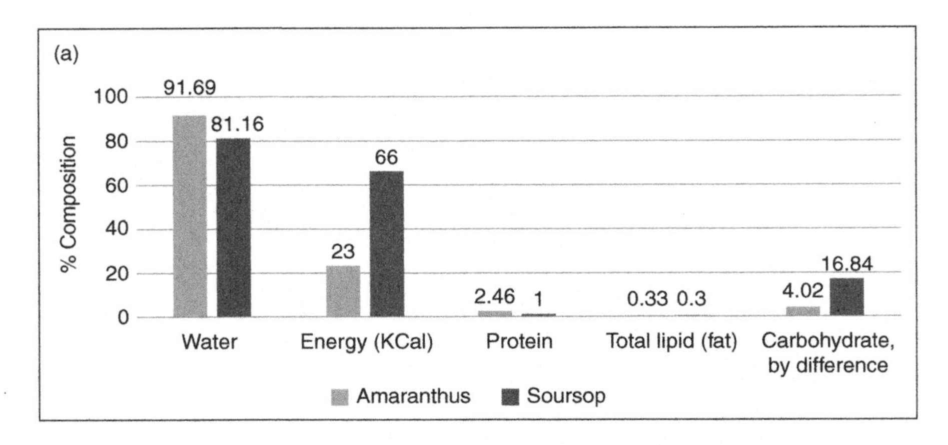

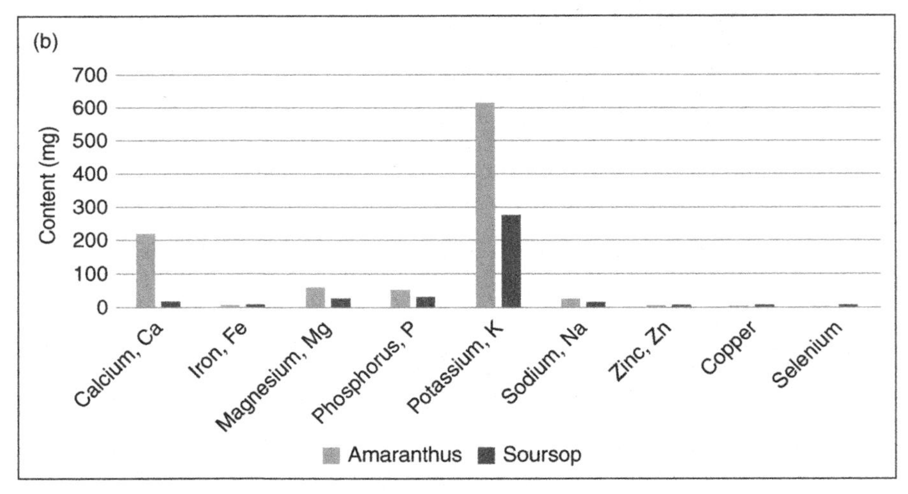

*Fig. 1.3. (a) Proximate composition and (b) mineral content of callaloo (*Amaranthus *sp.) and soursop (*A. muricata*). Selenium is in micrograms.* (Source: USDA National Nutrient Database, 2014.)

color when ripe, and typically the size of a chicken's egg (Figure 1.2b). The fruit has a sour but pleasant flavor when not fully mature that sweetens gradually as it ripens. It is eaten ripe or mature but still green with a variety of adjuvants, including salt and, in some cultures, a bit of red pepper (*Capsicum*). The June plum (*S. dulcis*) is eaten in the Americas as fruit and also used for making juice, but is also used as a vegetable in eastern Asia where it is made into a thick curry with coconut milk and spices. It is also preserved in the form of a chutney (like a thick semisweet/semisour jam), which is used to add a spicy kick to

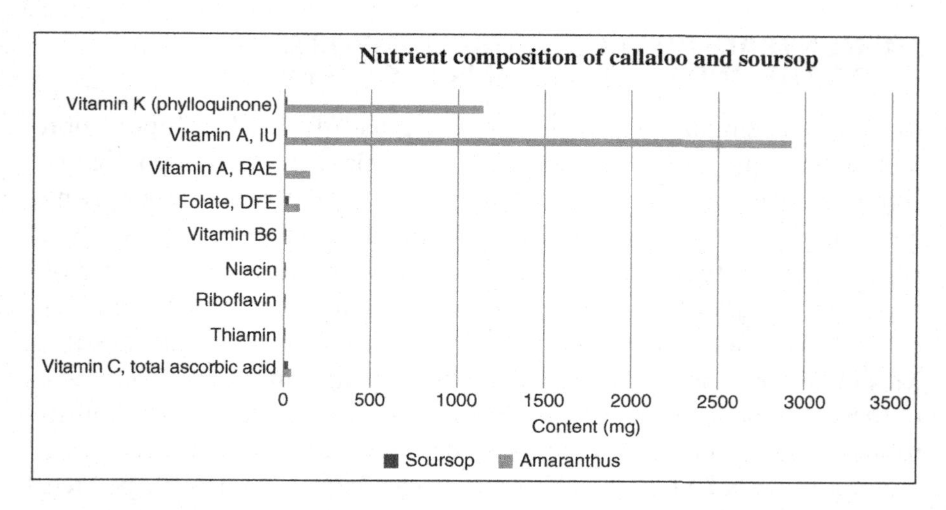

*Fig. 1.4. Vitamin content of callaloo (*Amaranthus *sp.) and soursop (*A. muricata*). Vitamins A and K are in international units (IU).* (Source: USDA National Nutrient Database, 2014.)

meals in the Caribbean. The fruit is rich in vitamin C, fiber, and minerals. It is in high demand in the United States, but supply is limited as many of the countries in which it is grown have endemic pests that are not allowed into the United States, resulting in their exclusion from the market. Countries that are able to overcome this challenge will find a growing market among Latin American, Caribbean, and Asian nationals and their descendants in the US market, as well as others who come to know this unique fruit.

The mamey sapote (*Pouteria sapota*), which looks similar to the fruit commonly called the mammee apple (*M. americana*) in the Caribbean, is another tropical fruit with growing demand in North America. This is because of its popularity in Mexico, in the rest of South and Central America, in the Caribbean, including the Dominican Republic and Cuba, and even in Florida where it is highly sought after by the large Hispanic community. The fruit has a high soluble fiber content and is said to help to reduce serum cholesterol, thereby promoting heart health, a characteristic which has contributed to its popularity. This, along with a variety of other tropical fruits, forms part of a growing list of what were previously "traditional" fruits from the Americas that will likely become a larger part of the increasingly diverse fruit choices of consumers in the US over the next 5 years.

1.4 NONTRADITIONAL VEGETABLE IMPORTS TO THE UNITED STATES: SPECIFIC EXAMPLES

In a manner similar to what is happening with fruits, US importation and consumption of nontraditional vegetables have increased significantly over the past 20 years, with developing countries now supplying more than 30% of the country's needs and more than 60% of its total vegetable imports (Hallam et al., 2004). Mexico is a major supplier of asparagus, eggplant, and onions, while Guatemala is a leading supplier of green peas and Kenya supplies 25% of world trade in green beans. Mexico alone supplies over 60% of US nontraditional vegetable needs. Cassava is supplied mainly from Central and South America, and mushrooms from China, as also are bok choy and other Asian vegetables. Among these NTAEs are ackees (Jamaica, Belize, and Haiti), peppers (including Scotch bonnet peppers – Figure 1.5b) and *Amaranthus*, sold as both the leafy vegetable – callaloo – and the seeds, the former coming mainly from the Caribbean (as callaloo), while the latter come from Central America, Mexico being the leading supplier. Several of these plants, including eggplant and ackees, will be discussed in more detail later on in this book. This short discussion will therefore focus on *Amaranthus*, a unique and multifaceted vegetable.

Hundreds of metric tons per annum of *Amaranthus* are exported globally, with India and Mexico being major global suppliers, and exports from India to the Netherlands, Canada, Australia, Germany, and the United States being delivered at prices ranging from US$0.86 to 2.08/kg in 2015, with the United States being a major market. Much of what is sold is in the form of the seeds, which contain significantly more protein per gram (about 30% more) than comparable grains such as rice, sorghum, and rye (United States Department of Agriculture (USDA), 2015). Typically, *Amaranthus* is a highly nutritive food that is a very good source of vitamins A and K and folate, as well as potassium, magnesium, calcium, and fiber, and is a good source of, for example, vitamin C, iron, and phosphorous (Figures 1.3a,b and 1.4). The product is gluten free and its amino acid profile is very good, making it an exceptional protein source. It contains all eight essential amino acids, which are typically absent from other grains. In fact, the World Health Organization (WHO) has recommended *Amaranthus* as a well-balanced food and NASA has also recommended its use as a food on space missions (Howard, 2013).

*Fig. 1.5. Two traditional developing-country exported vegetables: (a) callaloo (*Amaranthus *sp.) and (b) Scotch bonnet peppers (*Capsicum frutescens*).*

While much of the *Amaranthus* consumed is in the form of the seeds, there is also a growing import trade in the United States for callaloo (Figure 1.5a), the leafy vegetable of *A. dubius*, particularly in the wintertime when seasonal collard greens and other leafy greens are not readily available in the south. Like the seeds of the plant, cooked amaranth leaves are a good source of vitamin A, vitamin C, and folate; they are also a complementing source of other vitamins such as thiamine, niacin, and riboflavin, plus some dietary minerals including calcium, iron, potassium, zinc, copper, and manganese (United States Department of Agriculture (USDA), 2015; WebMD, 2015). Consequently, this nontraditional vegetable import that is now largely used by the immigrant population in the United States will become a part of the mainstream diet, once issues such as formulation, presentation, and culinary preparation options are addressed.

1.5 SANITARY AND PHYTOSANITARY/TECHNICAL BARRIERS TO TRADE ISSUES

The discussion on the growing export trade in NTAEs from emerging markets to developed countries could not be complete without a consideration of the issue of nontariff barriers to trade (NTBs). These are often seen by importing countries as required to protect their consumers or to ensure fair trade in all goods in their market, but as unreasonable by developing countries, who are often limited in their national scientific and technical infrastructure and personnel. This creates challenges in being able to do any research required; effectively implement control, food safety, quality, or national traceability systems; or provide the information requested and handle the delicate, often very technical, negotiations required. Below are summarized some of the SPS and TBT issues to be considered in dealing with market-access issues.

Among the more common SPS/TBT issues that produce importers and exporters have to handle are restrictions due to pest or disease concerns and requirements specifying certain postharvest treatment and fumigation that must be applied or avoided in order that the goods offered for import are compliant. Other SPS and TBT issues affecting various fruit and vegetable exports include:

- *Effective control of disease transmission*, e.g., fire blight, brown rot, canker, potato wart, fungus, and other unspecified diseases

- *Prevention of pest transmission*, e.g., codling moth, golden nematode, fruit flies, moths, and other unspecified quarantine pests
- *Chemical and pesticide residues*, e.g., methyl bromide, hydrogen gas; also maximum residual levels (MRLs) for certain pesticides
- *Treatment and mitigation requirements*, e.g., chemical and other treatment options, including fumigation and quarantine.

Exporters or importers are often required to provide evidence that these requirements have been met and, where they may not be applicable, technical information and data to support exemption from or removal of the requirements. For other products, issues of their safety, nutritional content, labeling, the use of the product in folk medicine, and the scientific basis for any claims made often arise.

In dealing with SPS/TBT issues, it is often best to engage the regulatory body (competent authority – CA) of the importing country to determine how best to ensure compliance with their requirements. Most are typically quite accommodating and will often work with the importer (and/or exporter and exporting country) to agree an approach that facilitates trade within the context of their legal requirements. An example of this is the approach taken by the USDA Animal and Plant Health Inspection Service (APHIS) with kiwifruit, for which there is high and growing demand in the United States. This is presented by reproducing in full the notice posted by the USDA APHIS in the Federal Register advising of the proposed approach to be taken to mitigate any risk to US crops from pests that may be imported along with the fruit (USDA, 2014). The posting reads:

Importation of Kiwi From Chile Into the United States: A Proposed Rule by the Animal and Plant Health Inspection Service on 10/16/2014

Summary

We are proposing to amend the fruits and vegetables regulations to list kiwi (*Actinidia deliciosa* and *Actinidia chinensis*) from Chile as eligible for importation into the United States subject to a systems approach. Under this systems approach, the fruit would have to be grown in a place of production that is registered with the Government of Chile and certified as having a low prevalence of *Brevipalpus chilensis*. The fruit would have to undergo preharvest sampling at the registered production site. Following postharvest processing, the fruit would have to be inspected in Chile at an approved inspection site. Each consignment of fruit would have to be accompanied by a phytosanitary certificate with an additional declaration stating that the fruit had been found free of *B. chilensis* based on field and packinghouse inspections. This proposed rule would allow for the safe importation of kiwi from Chile using mitigation measures other than fumigation with methyl bromide.

For the full rule see: https://www.federalregister.gov/articles/2014/10/16/2014-24631/importation-of-kiwi-from-chile-into-the-united-states

This proposed rule and, no doubt, the process that led to its promulgation, seeks to facilitate trade between the United States and an export partner where both have an interest in the trade occurring and where there are real risks to the phytosanitary health (in this case, from imported pests) of the importing country. A rule such as this would typically be the outcome of close collaboration between the CAs of both countries and the importing and exporting firms. This represents a viable option in dealing with SPS and TBT issues, and mirrors others that will be discussed later on in this book.

While actions such as the above ruling require bilateral action and cooperation, often the success of exporting entities lies within the firm or the sector. Industries such as Kenya's fresh produce sector have shown that well-organized industries in developing countries can use stricter standards and SPS requirements as a catalyst for change, and profit in the process if they are able to meet the requirements (Jaffee and Henson, 2005). There are many cases in which developing countries have faced restrictions because of their inability to meet food safety or agricultural health requirements. The outcome often depends on the nature of the relationships within the country involved and the industry; the strength of the CA and/or supporting public sector infrastructure in the exporting country; and the availability, depth of experience, and competence of industry, private sector, or academic personnel who will have to collectively solve the problem. In some cases, well-established export businesses and export-dependent sectors have been crippled or destroyed by the implementation of new, stricter standards, and this has had negative consequences for the livelihoods of industry stakeholders and the country. In other cases, countries have managed to gain or maintain access to high-value markets in developed countries, despite the tough technical requirements. The outcome, therefore, varies and, as noted by Jaffee and Henson (2005), the matter is not as clear cut as some commentators suggest. Whatever the outcome, the fact is that market-access requirements and import (regulatory) standards have become increasingly important as a determinant of the international competitiveness of developing-country firms and the countries themselves, especially for the export of high-value agricultural and food products. It is therefore incumbent on any firm or country that intends to be a successful participant in markets such as the US market for fruits and vegetables to arm itself with the knowledge that can facilitate its success.

This book will seek to provide exporters and trade and regulatory practitioners, particularly from developing countries, with the tools to deal head-on with SPS- and TBT-linked market-access challenges. It will do so by exploring the issues in detail and giving practical examples that can be used as guides to address SPS or TBT deficiencies. It will examine approaches to dealing with some of the most difficult SPS- and TBT-based market-access challenges using examples of successful approaches for key exports from Jamaica and the Caribbean region, with Jamaica's ackee being a prime example.

1.6 SUMMARY

Despite the challenges that have to be overcome for exporters from developing countries seeking to send traditional foods into developed-country markets, current world trends provide significant opportunities for growth in this area. As shown in this chapter, the demand in developed countries for traditional fruits and vegetables from developing countries is very favorable, including in the United States on which the focus is placed, and is expected to continue to be strong. This trend has proven to be very positive for the growth of the food industry in developing countries, and this has widened and deepened the options for food industry professionals. They have, in turn, further accelerated the development of the sector. This synergistic cycle has been and will remain central to the ability of emerging economies to actively participate in trade, and will prove even more important in the future as they have to continue to strengthen their ability to meet import requirements and deal with any sanitary and phytosanitary (SPS) and technical barrier to trade (TBT) issues that may arise.

CHAPTER 2

Natural Toxins in Fruits and Vegetables: *Blighia sapida* and Hypoglycin

Jose Jackson-Malete[1], Orane Blake[2], and André Gordon[3]

[1]Botswana Institute for Technology Research and Innovation (BITRI), Gaborone, Botswana
[2]A. T. Kearney Inc, Dallas, Texas, USA
[3]Technological Solutions Limited, Kingston, Jamaica

ABSTRACT

Many common foods that are widely consumed in temperate countries contain natural toxins. These include potatoes (*Solanum tuberosum*), which contain glycoalkaloids; rhubarb (*Rheum rhabarbarum*), which contains oxalic acid and anthraquinones; and eggplant (aubergine – *Solanum melongena*), which contains histamines. Consumption of these foods has not been restricted because consumers know how to handle them properly, resulting in a tolerable risk. Traditional fruits and vegetables from developing countries that also contain natural toxins, like cassava (*Manihot esculenta*), which contains cyanogenic glycosides, bok choy (*Brassica rapa*), which contains glucosinolates, and ackee (*Blighia sapida*), which contains hypoglycin A (HGA), are also now among food choices in developed countries. This chapter examines the issue of the food safety of plant-based foods that contain known natural toxins with a focus on traditional food from developing countries. It examines the issue of the toxicity of HGA, including its maximum tolerated dose (MTD) and median lethal dose (LD_{50}) in the context of other well-known toxicants in food

Food Safety and Quality Systems in Developing Countries. http://dx.doi.org/10.1016/B978-0-12-801227-7.00002-0

products. It reviews consumption and toxin-intake data and uses ackee as an example of how traditional foods from developing countries whose toxicological properties are known can be routinely consumed in a manner that ensures safety.

Keywords: developing countries; natural toxins; fruits and vegetables; *Solanum melongena*; *Blighia sapida*; hypoglycin A; MTD; LD_{50}

2.1 NATURAL TOXINS IN FRUITS AND VEGETABLES

Fresh vegetables and fruits contain nutrients that are essential for growth and health and are important components of a healthy diet. Their nutrient composition depends not only on botanical variety, cultivation practices, and weather, but also on the degree of maturity prior to harvest and the continuation of maturation post-harvest, which is influenced by storage conditions (Potter and Hotchkiss, 1995). Some vegetables and fruits, however, may also contain natural toxins, which are poisonous substances that are present naturally and are produced by plants to defend themselves against fungi, insects, and predators. These toxins offer a protective mechanism for the plant and while not harmful to the plant themselves, may potentially be harmful to human and animal health (Crews and Clarke, 2014). Cultures that use traditional foods have learnt to avoid those which contain naturally occurring acute toxicants or to process them to remove the toxicant (Potter and Hotchkiss, 1995). Nevertheless, there have been reports of food poisoning cases suspected to have been caused by consumption of raw vegetables and fruits containing natural toxins (Crews and Clarke, 2014), indicating that the matter requires attention, particularly as regards foods widely consumed or traded across borders.

The risk of poisoning by natural toxins in fruits and vegetables can normally be avoided or significantly reduced in traditional commerce by ensuring that the buyers have an understanding of the products being bought and taking care in the selection of the source from which the foods are procured. It is important to buy foods from reputable sources who can demonstrate a knowledge of the foods being sold, whether traditional or novel, as is currently now required of all food handlers under US, EU, and Canadian law, as well as that in many emerging economies. Informed exporters, importers, and consumers should also know that they should avoid buying or consuming products such as green potatoes or potatoes that are showing signs of sprouting or rotting. Likewise, consumers should know not to buy or consume raw or undercooked vegetables of uncertain origin if they are usually consumed cooked (FSANZ, 2014), as well as fruits that may contain natural toxicants and require special handling prior to consumption. Even with this knowledge, however, there are many plant-based foods that are routinely eaten that can be dangerous if not properly handled.

Eggplant (aubergine; *Solanum melongena*), also known as *brinjal* in Asia and South Africa, boulanger and melongene in the Caribbean, and elsewhere as guinea squash and garden egg, is commonly used in many cuisines around the world, particularly by vegetarians. This plant (Figure 2.1a), which is widely consumed in India and China, is known to cause itchy skin or mouth (oral allergy syndrome), among other symptoms, due to the presence of *histamines* in the plant. Thoroughly cooking eggplant prevents reactions in susceptible individuals, but at least one of the allergenic proteins survives the cooking process. Nutmeg – an important seasoning and flavoring in the meat, prepared foods, and hospitality industries – contains *myristicin*, which is toxic at high levels, causing headaches, nausea, dizziness, tachycardia, memory lapses, and hallucination, among other symptoms. Rhubarb (*Rheum rhabarbarum*), another widely consumed vegetable, and sorrel (*Rumex acetosa*), popular in tropical countries as a beverage, both contain *oxalic acid*, which is harmless in small amounts but can be fatal if consumed in large quantities. Rhubarb also contains *anthraquinones*. Both rhubarb and sorrel can cause severe poisoning and kidney damage that may be fatal (Canadian Food Inspection Agency (CFIA), 2012). Cabbage (*Brassica oleracea capitata*), a basic part of the diet in developed countries, and bok choy (*Brassica rapa* subspecies *pekinensis* and *chinensis*), also known as Chinese cabbage and heavily consumed in Asia and Latin America (Figure 2.1b), both contain *glucosinolates*. These compounds, while generally regarded as desirable for their anticancer properties, can also be toxic at high levels because of their goitrogenicity.

Beans such as green beans and red and white kidney beans, as well as cassava and bamboo shoots, need to be cooked thoroughly at boiling temperature after thorough soaking them in clean water to avoid

Fig. 2.1. Selected traditional foods: (a) eggplant, (b) bok choy, (c) sorrel.

potentially negative health consequences. Raw or inadequately cooked beans should not be used in the preparation of salad dishes. Fresh fruits, such as apples, apricots, and pears, are an important and nutritious part of a balanced diet; however, eating the seeds of these fruits should be avoided. Therefore, while many foods add variety, nutrition, and other health benefits to the diet, the preparation and consumption of those that contain natural toxins is an area that requires attention, an observation also made by the Canadian Food Inspection Agency (CFIA) (2012). Selected plants and their natural toxins are indicated in Table 2.1.

Although many foods contain toxins as naturally occurring constituents that can cause illness or adverse reactions (Dolan et al., 2010), the incidence of negative consequences from the consumption of these foods is relatively low. For the glycoalkaloids present in potatoes, the average daily intake varies from 14 mg or 0.28 mg/kg body weight (BW) in the United Kingdom to a high of 1.2 mg/kg BW in Sweden (Hopkins, 1994). This latter is above the dose (1 mg/kg BW) at which toxic effects such as nausea and diarrhea are observed, and close to the estimated median lethal dose (LD_{50}) of 3–6 mg/kg BW (BIBRA, 1995). Despite this, no one has suggested that restrictions should be imposed on the consumption of potatoes. The low incidence of adverse effects is the result of pragmatic solutions by regulatory agencies through the creative use of specifications, action levels, tolerances, warning labels, and prohibitions. Industry has also played a role by setting limits on certain substances and developing mitigation procedures for certain toxins (Dolan et al., 2010), inclusive of codifying and implementing industry best practices and other codes of practice. Nevertheless, although the risk of poisoning due to the consumption of food toxins is fairly low, there is always the possibility of illness arising from contamination, overconsumption, allergy, or an unpredictable idiosyncratic response. The purpose of the ensuing discussion, therefore, is to provide a template for how scientists, regulators, producers, exporters, and importers can approach the characterization of traditional plant-based foods containing natural toxins in a manner that meets the technical and scientific requirements for establishing safe handling and consumption practices, where these may be absent. The work done with ackee (*Blighia sapida*), the national fruit of Jamaica, is discussed, including a toxicological overview of hypoglycin, the natural toxin present in the ackee fruit.

Table 2.1. Selected Foods of Vegetable Origin and Their Natural Toxins

Fruit or Vegetable	Natural Toxin	Effects and Ways to Control the Toxin
Beans (such as green beans (*Phaseolus coccineus*), red kidney beans, and white kidney beans (*Phaseolus vulgaris*))	Phytohemagglutinin	Food poisoning caused by this toxin in raw and inadequately cooked beans has a short onset time (1–3 h), with symptoms of nausea, vomiting, and diarrhea. This toxin, a lectin, can be eliminated by thoroughly soaking and cooking the beans in boiling water. Canned beans that are retorted* are safe to eat without further cooking.
Cassava (*Manihot esculenta*) and bamboo shoots (*Bambusa vulgaris*)	Cyanogenic glycoside	The bitter type of cassava and fresh bamboo shoots have high levels of toxins. When inadequately cooked cassava or bamboo shoots are eaten, the toxin is transformed into hydrogen cyanide, which may result in food poisoning. Symptoms of cyanide poisoning occur within a few minutes and may include constriction of the throat, nausea, vomiting, and headache. Death has been reported in severe cases. Fresh bamboo shoots should be sliced into smaller pieces and cooked thoroughly. Cassava should be properly soaked and pressed to remove the cyanogenic glycoside.
Potatoes (*Solanum tuberosum*)	Glycoalkaloid (solanine glycoalkaloids: solanine and chaconine)	While the levels of glycoalkaloids are low in most potatoes, those that show signs of greening, or sprouting, or are physically damaged or rotting, may contain high levels. The majority of the toxin is present in the green area, the peel, or just below the peel of the potatoes. Glycoalkaloids have a bitter taste. Symptoms of poisoning may include a burning sensation in the mouth, severe stomachache, nausea, and vomiting. Potatoes should be stored in a dark, cool and dry place to minimize this risk. Cooking or frying cannot destroy the toxin, once formed.
Parsnips (*Pastinaca sativa*)	Furocoumarins	Produced as a way of protecting the plant when it has been stressed. The concentration of the toxin is usually highest in the peel or surface layer of the plant or around any damaged areas. Furocoumarin toxins can cause stomachache. It is important to peel the parsnip before cooking and remove any damaged parts. The levels of toxin drop when the parsnip is cooked by baking, microwaving, or boiling. Discard water used for cooking.
Zucchini (*Cucurbita pepo*)	Cucurbitacins	These toxins give zucchini a bitter taste. This is rarely found in commercially grown zucchini. Bitter zucchinis can cause vomiting, stomach cramps, diarrhea, and collapse. Avoid zucchinis that have a strong, unpleasant smell or taste bitter.
Eggplant (*S. melongena*)	Histamines	These toxins cause mild headaches, stomach upset, and itchy skin or mouth. Up to just below 10% of Indians have been found to show allergic symptoms after consuming aubergine (eggplant). Hypersensitive (atopic) individuals are more likely to have a reaction to eggplant, possibly because it is high in histamines. While at least one of the allergenic proteins is thermostable, cooking appears to prevent allergic reactions in some people.
Ackee (*B. sapida*)	Hypoglycin	Immature ackee fruit contains hypoglycin that can cause serious health effects owing to hypoglycemia. Only the yellow fleshy arillus (aril) from a fully mature, opened fruit is edible. The shiny black seeds, the raphe, and the pods (Figure 2.2) are inedible and toxic. The fruit is poisonous unless ripe and having opened naturally on the tree.

**Cooked in a hermetically sealed container using steam at 121°C under pressure.*

2.2 THE ACKEE FRUIT

The ackee (*B. sapida*) is a member of the soapberry family (Sapindaceae) and is native to West Africa, with Cote d'Ivoire, Benin, Burkina Faso, Ghana, Nigeria, and Mali being among the countries in which it can be found. The plant was named *B. sapida* by Koenig in 1806; however, the colloquial name ackee was derived from the terms *ankye* and *akye-fufuo*, used to describe the tree and its fruit in Twi, the chief language of the Republic of Ghana (Irvine, 1930). It is also known as vegetable brain, achee, akee apple, and akee. The fruit was brought to the Americas in 1778 by Dr. Thomas Clarke, Jamaica's first botanist, who had obtained the seeds from a West African slave ship and planted it on the island (Broughton, 1794). Since then, it has been introduced to several other countries in the Americas, including the Caribbean countries of St. Vincent, Grenada, St. Lucia, Belize, and Haiti. It is also grown in Florida, a popular destination for exports of the fruit.

The fruit consists of a red to yellow inedible outer pod (the husk) that protects the growing seed and edible component of the fruit when immature. When fully ripened, the pod splits open to reveal two to four shiny black seeds, each having a cream-colored fleshy lobe, the arillus (aril), at its base (Figure 2.2). Traditionally, in both West Africa and the Americas, it is when the fruit has opened naturally on the tree that the ackees are harvested and the arils removed and cleaned, in preparation for cooking. The aril of the ackee fruit is typically about 4–6 cm long, depending on the variety, country, and source, and weighs about 75–100 g (Gordon, 1995a). This is the part of the fruit that is eaten, with

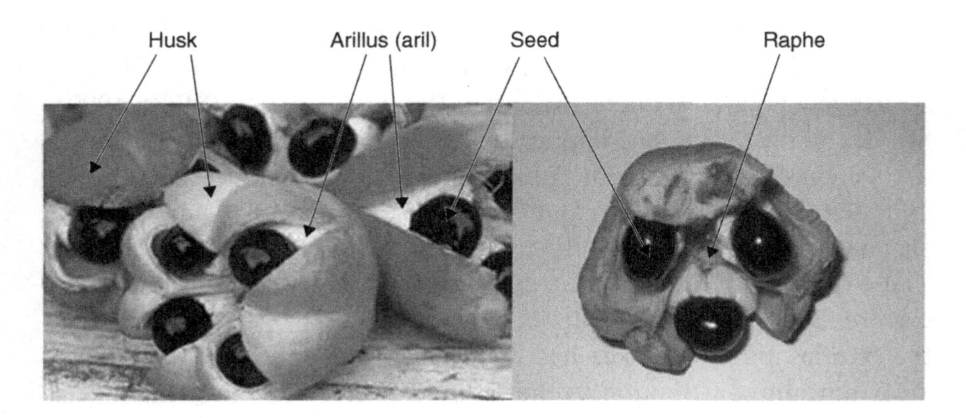

Fig. 2.2. The ackee fruit.

Table 2.2. Nutritional Composition of Fresh and Canned Ackee

Nutrient	Content/100 g Fresh Weight	Content/100 g Canned, Drained Aril
Calcium (mg)	98	35
Iron (mg)	NA	0.7
Potassium (mg)	NA	270
Sodium (mg)	NA	240
Zinc (mg)	NA	1
Carotene (mg)	0.1	NA
Thiamine (mg)	0.18	0.03
Riboflavin (mg)	3.74	0.07
Niacin (mg)	65	1.1
Folacin (total) (μg)	NA	41
Cyanocobalamin (μg)	NA	0
Vitamin C (mg)	NA	30
Water (g)	57.6	76
Energy (kcal)	NA	151
Protein (g)	8.75	2.9
Fat (total) (g)	3.45	15.2
Saturated fat (g)	NA	0
Cholesterol (mg)	NA	0
Carbohydrates (total) (g)	1.87	0.8
Dietary fiber (g)	9.55	2.7

NA – not analysed "Fresh" means "raw" and "uncooked"
Source: Caribbean Food and Nutrition Institute (CFNI), 1998; http://www.foodreference.com/html/artackee.html

other parts of the fruit and the tree being traditionally used for other purposes (discussed in Chapters 3 and 5).

The nutritional composition of ackee as measured in the aril is shown in Table 2.2. The fruit has a high water content (~77%) and, at 15.2%, a relatively high fat content (Scientific Research Council (SRC), 1999). Studies on the fatty acid composition of the aril show that 51–58% of its dry weight consists of lipids (Emanuel et al., 2013). Linoleic, palmitic, and stearic acids are the major fatty acids present, with linoleic acid accounting for over 55% of the total fatty acids (Odutuga et al., 1992). These results indicate that the fruit is of significant nutritive value (Lancashire, 2008). They also support subsequent work that showed that the purified oil from ackee has high nutritional value and may make an important contribution to the fatty acid intake of many Jamaicans (Odutuga et al., 1992) and those eating it in western Africa.

Ackees have been prepared and eaten in western Africa for many years, particularly in northern Cote d'Ivoire in the Katiola, Sinématiali, and Khorogho regions, typically as a dried paste, and in the Toussiana and Peni divisions in Burkina Faso. In Jamaica, the consumption of the fruit dates back to the nineteenth century (Ashurst, 1971; Kean, 1989). As the national fruit of Jamaica, ackee forms an integral part of the Jamaican diet and is also a standard breakfast menu item at Jamaican hotels where, over the years, it has been consumed by millions of visitors to the island. Ackee is typically combined with a savory meat or other vegetables and had for breakfast or as part of the main daily meal. The arils could also be cooked with codfish, onions, and tomatoes, or added to stewed beef and salted pork with a range of herbs and spices, or they could be curried and eaten with rice (Emanuel and Benkeblia, 2011). The fruit is also eaten in several other countries, largely in areas populated by Jamaicans and Caribbean nationals, and at Caribbean restaurants throughout North America and the United Kingdom that feature it on their menu. Ackees are a major contributor to jobs in the agro-processing sector in Jamaica and are the backbone of the business of several small agro processors and rural communities (Gordon, 1995b). With exports of approximately US$350,000, US$600,000, and US$300,000, respectively, in 2013 from Belize, Haiti, and Cote d'Ivoire, and exports from Jamaica increasing from US$4.4 million in 1999 (Gordon, 1999a) to US$15.3 million in 2013 (STATIN, 2014), ackees are also an important contributor to these countries' export earnings from nontraditional agricultural exports (International Trade Center (ITC), 2013). The cultural, dietary, and economic significance of this fruit (as for any other widely consumed fruit or vegetable that also contains a natural toxin) requires that a thorough understanding of the toxin, its management, and the risks from its consumption is developed to facilitate its trade in the international marketplace.

2.3 HYPOGLYCIN (HGA), THE NATURAL TOXIN IN ACKEE

Since the nineteenth century, it had been recognized that the immature (unripe) ackee fruit may be poisonous (Bowery, 1892). It was postulated that glycosides and saponins were the toxic components in the fruit (Evans and Arnold, 1938). However, these suggestions were not supported when two groups independently isolated two toxic compounds

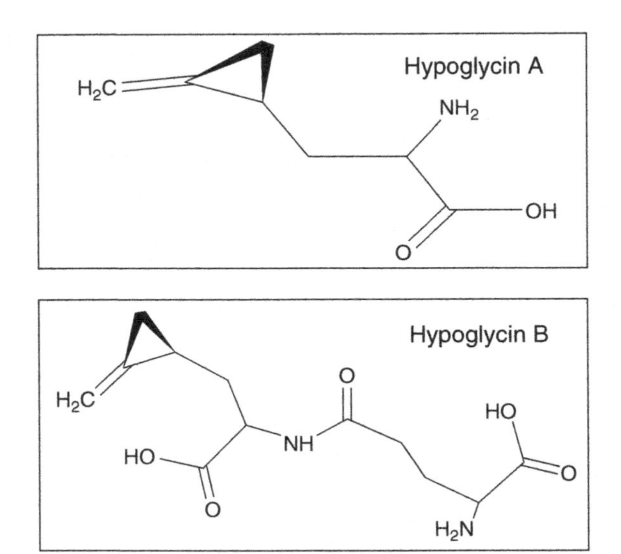

Fig. 2.3. Schematic structure of the hypoglycin A and B toxins.

from the unripe fruit, and later called them hypoglycin A and B (Hassall et al., 1954; Hassall and Reyle, 1955; Holt and Leppla, 1956). The chemical structures of these molecules were later elucidated and are illustrated in schematic form in Figure 2.3. Subsequent investigation revealed that hypoglycin A (HGA) exists as a mixture of (2S,4R) and (2S,4S) diastereoisomers at the C4 position, with the (2S,4R) isomer being 17% in excess of the (2S,4S) isomer (Baldwin et al., 1990, 1994).

HGA is typically present at levels in excess of 1000 ppm in the immature fruit (Brown et al., 1992; Gordon and Jackson, 2013). At these high levels, consumption of the immature fruit produces an acute toxic effect that usually occurs within 6–48 h, with symptoms including vomiting, drowsiness, and hypoglycemia (Manchester, 1974; Morton, 1987). Coma and death can occur within 12 h in severe cases of toxicity (Hassall and Reyle, 1955). This illness, which has historically had a much higher incidence among young children than adults (Hill, 1952; Kean, 1989; Scott, 1917), is commonly known as Jamaican vomiting sickness (JVS). The toxicity is due to the metabolization of HGA in a manner similar to that of branched-chain amino acids (Figure 2.4), producing the active metabolite methylenecyclopropylacetyl coenzyme A (MCPA-CoA), which contributes to the etiology of JVS (Billington et al., 1978; Tanaka, 1975;

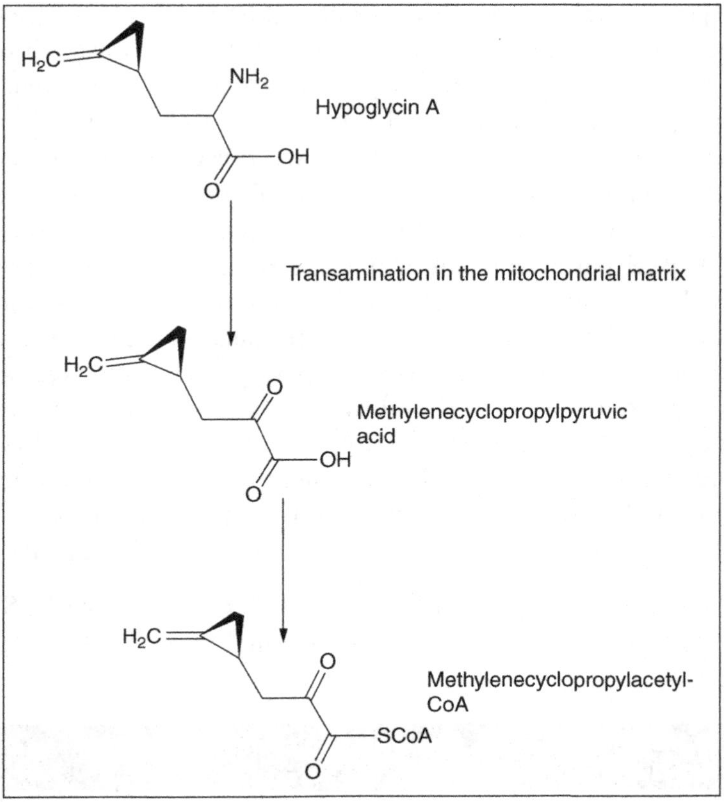

H_2C NH_2 Hypoglycin A OH O

Transamination in the mitochondrial matrix

H_2C O Methylenecyclopropylpyruvic acid OH O

H_2C O Methylenecyclopropylacetyl-CoA SCoA O

Fig. 2.4. Conversion of hypoglycin A to MCPA-CoA.

von Holt, 1966). Consumption of mature (ripe) ackee, which contains 100 ppm or less HGA, is harmless (Brown et al., 1992; Whitaker et al., 2007). The toxic effects, the mechanism of toxicity, the details of the illness, and its impact will be discussed in greater detail in Chapter 4.

2.4 DIETARY EXPOSURE TO THE NATURAL TOXIN HYPOGLYCIN

In order to understand the risk associated with the toxin, HGA, and its intake by eating the ackee, it is necessary to investigate typical levels of dietary exposure among the consuming public. Dietary exposure to HGA was reported by Blake et al. (2004) in children and adults, using ackee consumption data and quantifying HGA levels in typical ackee diets. Blake et al. (2004) evaluated the mean and maximum HGA intake in Jamaica with respect to age, socioeconomic status (SES), geographical location (rural and urban), and gender. They found that ackee consumption was highest in the lower socioeconomic group, particularly in children. HGA levels in typical ackee diets ranged from 1.21 to 89.28 μg HGA/g ackee (Blake et al., 2004). This wide range was likely due to the varying preparation and cooking methods used by individuals depending on their location and regional culinary subculture. Cultural practices suggest that the level of hypoglycin in the fruit is affected by the efficiency of cleaning, particularly in removing the raphe, in addition to the time for which the product is boiled in water. This is often dependent on the variety and maturity of the ackee and the individual preparing the meal. Ackee varieties differ in texture (Bowen, 2006), which affects the boiling time required to develop optimal texture and flavor. Since HGA is water soluble, the traditional view is that varying the boiling times will affect its concentration in the final product. Using ackees with a range of maturities will also naturally affect the HGA content of the final product (Blake et al., 2004).

The mean and maximum HGA intake with respect to age, SES, rural and urban location, and gender are shown in Table 2.3. The data show

Table 2.3. Dietary Intake of HGA by Jamaican Consumers of Different Ages, SES, Geographical Location and Gender

Individual	Mean Intake HG (μg/day/kg BW)	Max. Intake HG (μg/day/kg BW)
Age		
Children	8.18 ± 6.04 a	115.2
Adolescents	7.41 ± 3.52 b	79.3
Adults	6.55 ± 3.12 c	69.9
Elderly	6.58 ± 4.06 c	66.95
SES		
Upper SES	4.40 ± 3.01 a	51.79
Middle SES	4.40 ± 3.01 a	51.79
Lower SES	7.87 ± 6.48 b	105.36
Location		
Kingston/St. Andrew	5.10 ± 4.17 a	67.86
St. Thomas	7.87 ± 6.25 b	102.68
Gender		
Male	8.34 ± 6.25 a	104.46
Female	6.72 ± 5.56 b	90.18

that children consumed a greater quantity of HGA than adolescents, the elderly, or adults. Individuals from the lower SES consumed more HGA than individuals from the upper and middle SES. Similarly, individuals from St. Thomas (a rural location) consumed a much larger quantity than individuals from Kingston and St. Andrew (an urban location). Males also consumed a greater quantity of HGA than did females. This suggests that males would be more likely to have a higher incidence of ackee toxicity than females, should the fruit be immature (unfit for consumption).

This study indicated that consumers in Jamaica are exposed to varying levels of the toxin depending on their age, SES, location, and gender. Should the fruit be immature and therefore toxic, children appeared to be more at risk than other members of a family, as they consume far greater quantities relative to their body weight. Individuals from the lower SES are also at greater risk of experiencing ackee toxicity since they also consume a far greater quantity of ackee and HGA than individuals in the middle or upper SES. Similarly, persons from rural locations and males are also at greater risk. These findings explain the observations in the literature in the period up to 2000 that, in the few cases where there have been outbreaks in Jamaica, most of the reported cases of ackee toxicity have involved young male children from the lower SES in rural areas (CDC, 1992; Hill, 1952).

2.5 ACUTE AND CHRONIC TOXICITY

The acute and chronic toxicities of natural toxins in foods are important as these influence the choice of the most effective approaches to prevent poisoning from the use of these foods. In order to advance the safe trade in and use of these foods, scientifically sound data-based practices need to be developed for all such foods. The impact of toxicity was, for some time, assessed by the LD_{50} of toxins, which indicated the dose level at which half of the test population would not survive. In recent times, this has been seen as unethical and its usefulness questioned because it does not consider the more likely and important sublethal effects of the toxin. Current practice uses the no toxic effect level (NTEL), no observed effect level (NOEL), and the maximum tolerated dose (MTD). In reviewing studies on toxicity, therefore, cognizance needs to be taken of the differences in the approach

Table 2.4. Lethal Doses (LD_{50}) for Hypoglycin

Animal	LD_{50} (mg/kg BW)
Rats	98
Mice	160 (min)
Monkeys	20
Rabbits	10
Humans	NE

NE – not evaluated

that would have been used when the foundation studies were done, as against what would now be regarded as ethically sound and useful.

For ackee (*B. sapida*), there have been several studies that have sought to establish the levels of HGA that are associated with acute toxic effects and also the intake levels that cause chronic toxicity. The LD_{50} of HGA for animals other than humans has been examined by several investigators (Blake et al., 2006; Brooks and Audretsch, 1971; Senior and Sheratt, 1968; Singh et al., 1992; Tanaka et al., 1972; von Holt et al., 1966). They found that the LD_{50} of HGA for animals ranges from 10 mg/kg BW for rabbits to greater than 160 mg/kg BW for mice (Table 2.4). Rats have an LD_{50} of 98 mg/kg BW, while hamsters are much less sensitive with an LD_{50} of greater than 150 mg/kg BW (BIBRA, 1995). Blake et al. (2006) found that the acute toxic dose of HGA for male and female rats was 231.19 ± 62.55 and 215.99 ± 63.33 mg/kg BW, respectively. This was considerably greater than the LD_{50} of just under 100 mg/kg BW reported in a study when aqueous HGA was administered orally (BIBRA, 1995).

In their study, Blake et al. (2006) used ackees as the medium for HGA administration in laboratory rats. HGA toxicity was determined by feeding male and female Sprague–Dawley rats a control diet and ackee diets that contained 4–3840 ppm of HGA. The fixed-dose method was used to quantify the acute toxic dose of HGA and was determined by feeding a diet consisting of the lowest HGA concentration; this was increased to the next highest dose after 24 h, until toxicity was observed. The MTD of HGA was determined by feeding rats the ackee and control diets over a 30-day period. The MTD of HGA in both male and female rats in this study was 1.50 ± 0.07 mg/kg BW/day. This contrasts with previous studies of the effects of repeated exposure to the toxin, where up to 20 mg/kg BW daily for 30 days showed no marked effects on rats (BIBRA, 1995; Brooks and Audretsch, 1971).

In considering the results of earlier toxicological studies, cognizance must be taken of the fact that the HGA being used was likely contaminated with leucine, isoleucine, and, possibly, valine (Sarwar and Botting, 1994; Sarwar et al., 1988) although this would not have been known to the researchers involved. The presence of the other amino acids may have skewed the results obtained as the HGA being used was not pure. Additionally, most of these studies were conducted using intraperitoneal or intravenous injection, although the differences between oral administration and these two other means have been shown to be minimal in most cases (BIBRA, 1995). Taken together, the findings of these studies suggest that tolerance to HGA is greater when it is consumed in ackees (Blake et al., 2006), rather than taken directly orally or administered intraperitoneally. The difference observed was probably due to a reduced absorption of HGA in the gut of the rats, because of the matrix in which it was administered orally (the ackee), thus requiring ingestion of a larger amount of the HGA to produce acute toxicity. Therefore, the form in which HGA is administered could affect the toxicological properties it exhibits. Blake et al. (2006) concluded that since the ackee is the normal route for HGA into the body, then that mode of administration is likely to provide a more reliable estimate of HGA toxicity for hazard assessments.

The MTDs reported are more than 150 times the daily intake reported for the highest consumers of ackee in Jamaica (Blake et al., 2004), indicating that substantial abuse of the handling of the fruit and consumption of excessively high levels of immature fruit would be required to occasion the symptoms of acute or chronic toxicity of HGA poisoning. This is also supported by the findings of a study by Professor Robert Bates of the University of Florida (Bates, 1991) and other sources (BIBRA, 1995). It should be noted, however, that HGA toxicity has been shown to double under conditions of starvation, hence the increased susceptibility of the malnourished to HGA poisoning. Nevertheless, toxicological studies to date indicate that the risk of HGA poisoning is low if proper handling of the fruit is undertaken.

2.6 CONCLUSIONS

In examining the issue of the toxicity of plant-based foods, cognizance has to be taken of the fact that many common such foods contain known toxins. These include potatoes and the glycoalkaloids they contain, as well as foods such as cassava that contain cyanogenic glycosides. In both these and other instances, there has been no public attempt to restrict the consumption of these foods because consumers are generally aware of how to handle them properly, resulting in a risk that is tolerable. Ackee, a prime example of a novel but traditional food, has also been a part of the diet of Jamaicans for many years, the majority of whom are aware of its toxicological properties and know how to prepare it to ensure safety. The issue of the toxicity of HGA from ackees should be examined in the same context as for other well-known toxicants in food products that are widely consumed and recognized as safe when handled as recommended. Characterizing and studying this plant and its toxin have reduced the concerns about its safety, as with other plants containing natural toxins. Proper handling and education of consumers will further minimize any potentially negative consequences of including this fruit as a part of the diet.

CHAPTER 3

The Life Cycle of Ackee (*Blighia sapida*): Environmental and Other Influences on Toxicity

André Gordon[1] and Jose Jackson-Malete[2]

[1]Technological Solutions Limited, Kingston, Jamaica
[2]Botswana Institute for Technology Research and Innovation (BITRI), Gaborone, Botswana

ABSTRACT

Fruits and vegetables from developing countries that have been a part of traditional diets and are now entering international commerce but contain natural toxins need to be properly characterized. It is important to understand where the toxin is located in the fruit, what influences the level of the toxin, and the mechanism of natural detoxification, as well as the factors that cause increased or reduced toxicity. This will allow for science-based management of the handling and preparation of food, to reduce or eliminate the risk of adverse health consequences. This chapter explores this issue and uses the ackee (*Blighia sapida*) as an example. It discusses changes in the toxin present in ackee hypoglycin A (HGA), with maturity, the research into the mechanism of detoxification of HGA, and the impact of geographical and varietal differences. It establishes that much of the traditional knowledge has a sound scientific base and outlines the relationship of this information to handling and

Food Safety and Quality Systems in Developing Countries. http://dx.doi.org/10.1016/B978-0-12-801227-7.00003-2

reaping practices which may influence the normal cycle of detoxification. The chapter also discusses how ackees must be properly controlled if this fruit is to be safely and successfully exploited commercially, which is also true for other horticultural products containing natural toxins.

Keywords: ackee; hypoglycin A; level of maturity; mechanism of detoxification; traditional knowledge; reaping practices; varietal differences

3.1 THE AGRONOMY OF ACKEE

Ackee, *Blighia sapida* Koenig (from the Sapindaceae family), is a polygamous, evergreen tree that originated in the forests of West Africa (Figure 3.1a). It is related to the lychee and the longan, grows to a height of up to 18 m in the wild, and has pinnate leaves composed of three to five pairs of ovoid leaflets with dimensions of 2–9 × 4–18 cm. It has auxiliary racemes with flowers that are small, greenish-white (Figure 3.1b), fragrant, about 5 mm in length, and borne in sprays (Brown, 1989; Henry, 1994). Commercial propagation is by seed or cuttings. Propagation by seeds results in the trees normally coming into bearing 3–4 years after planting, while propagation by tip cuttings results in lower-growing, spreading trees that bear fruit within 1–2 years (Senior, 1983). Once planted, the tree can grow to a height of 8–15 m, producing a thick green pod (Scientific Research Council, 1999), which upon maturation becomes red or yellow (Figure 3.2a). The fruit is borne in hanging clusters consisting of thick, leathery, greenish-yellow capsules (pods), which are ovoid and lobate in shape and are typically 3 × 6 cm long in the Caribbean and Florida, although the predominant varieties in some parts of West Africa are smaller (Figure 3.1a). The fruit normally consists of three lobes, although fruits with two or four are not unusual. The pods turn red when mature and split to expose fleshy, pale yellow or cream-colored arilli (Figure 3.2b), each with a hard, shiny, round black seed at its tip (Morton, 1987). Traditionally, it is at this time that the ackees are harvested and the arils removed and cleaned, in preparation for cooking.

Ackee trees were introduced into Jamaica in the eighteenth century by Dr. Sir Thomas Clarke (Bressler et al., 1969; Lewis, 1961), as mentioned in Chapter 2. The first ackees were taken to London in 1806 by Captain William Bligh (of "Mutiny on the Bounty" fame) on the HMS *Bounty*, hence the name *Blighia*. The term "sapida" refers to the savory

Fig. 3.1. (a) Ackee trees in an orchard in Côte d'Ivoire; (b) ackee on a tree in Florida, with flowering blossoms. (Source: A. Gordon, 2015.)

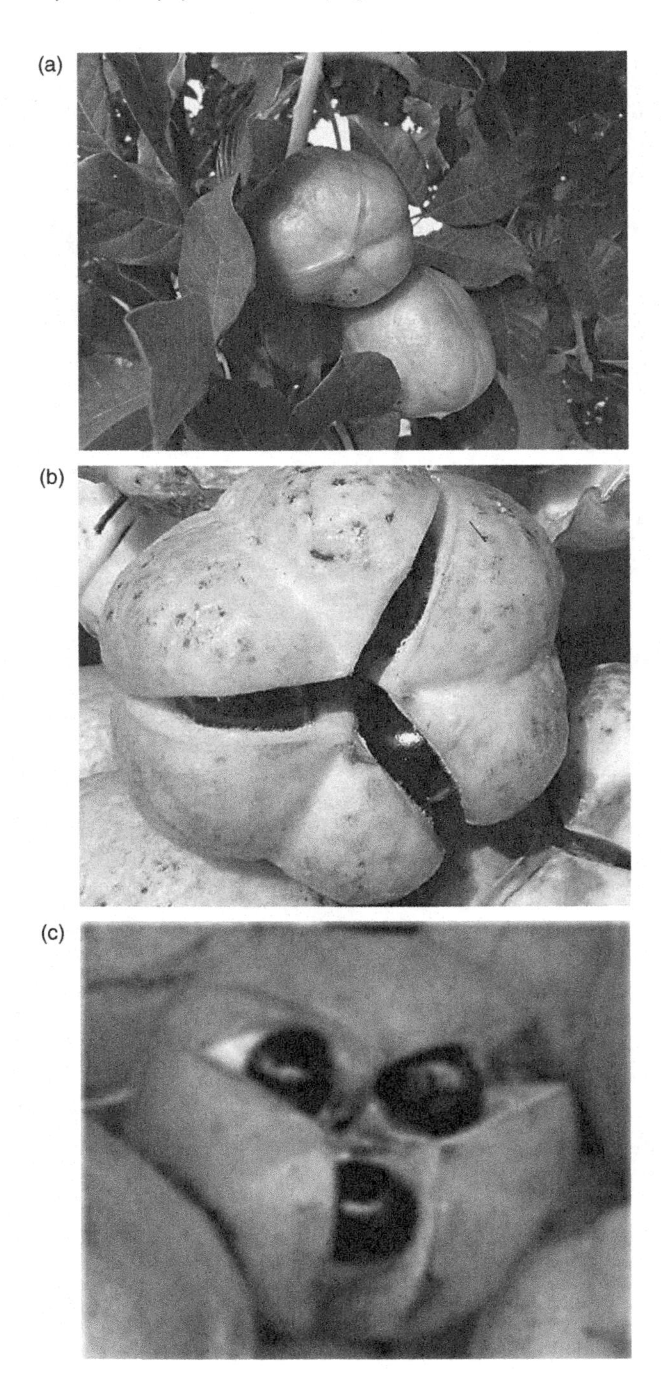

Fig. 3.2. The stages of maturity of the ackee fruit: (a) stage 5 (mature but unopened), (b) stage 6, (c) stage 7, (d) stage 8, (e) stage 9, (f) stage 10. (Source: A. Gordon, 2015.)

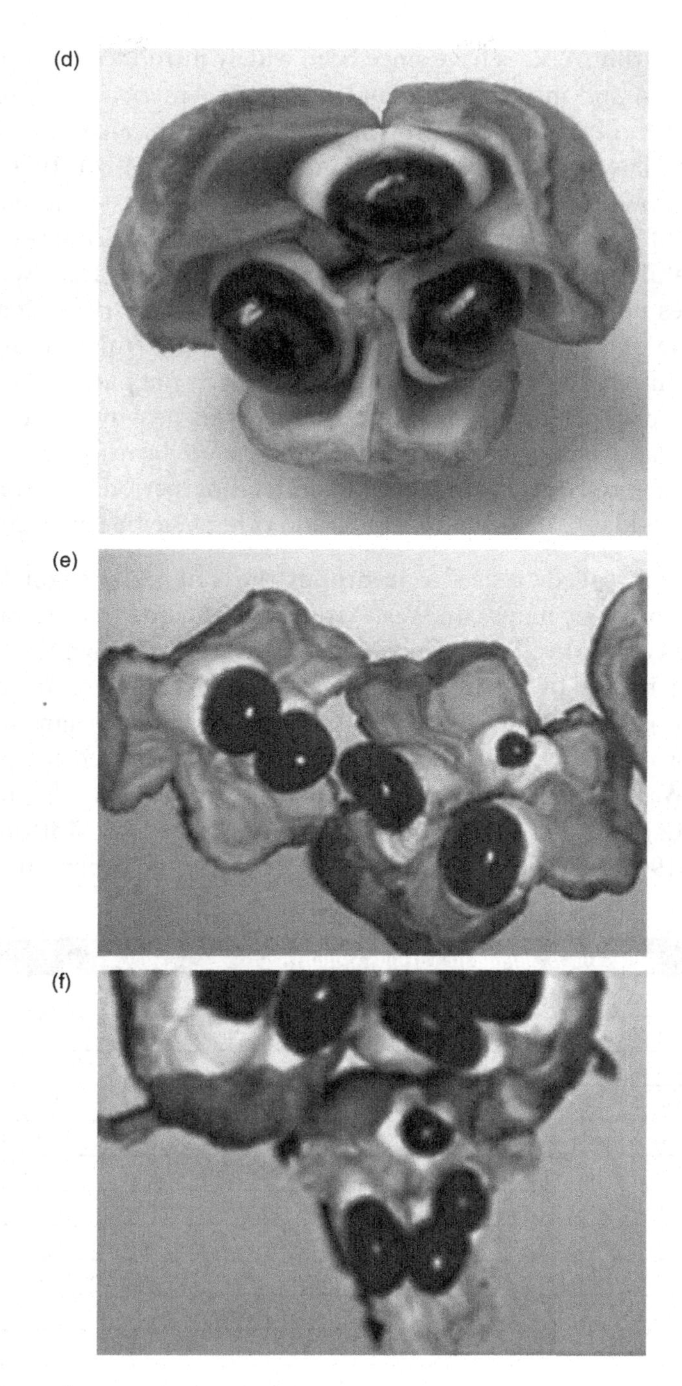

Fig. 3.2. (Continued)

taste of the fruit. Ackees have since been widely introduced into Florida (Figure 3.1b) and into several countries in the Americas by Jamaicans resident in those countries (Brown, 1989). They are also cultivated in India and Hawaii (Lewis, 1961; Perkins and Payne, 1978). In Jamaica, the trees grow in all types of soils up to 600 m above sea level and in locations of heavy rainfall. However, they thrive best in areas of medium rainfall and fruit best at low altitudes; although trees will grow at altitudes up to 850 m, they will not flower (Brown, 1989). The trees are capable of bearing all year round, but typically fruited most heavily from January to March and June to August in the past. This pattern has changed in recent years, and trees now fruit heavily in October to November and January to March, although some bearing often occurs in midAugust to September. These main fruiting periods are similar to those in Florida, Côte d'Ivoire, Haiti, and other Caribbean countries.

Commonly called "ackee" for centuries, the fruit is known on different continents by many names. In West Africa, the fruit and tree are variously called "ankye," "akye fufo," "akye-fufuo," or "isin" (Brown, 1989). The fruit is also known by a variety of other names, depending on the country in which it is found (Table 3.1). The tree is regarded as ornamental and is used for making furniture, with its blossoms being used for perfume in some West African countries. Oils derived from the seeds and husk (dried pods) are used for making soaps (Hawkes, 1972; Morton, 1987; Seaforth, 1962). The dried seeds, fruit bark, leaves, and other parts of the

Table 3.1. Names for *B. sapida* in Various Countries

Names	**Country**
Huevo vegetal; fruto de huevo	Guatemala, Panama
Arbol del huevo; pera roja	Mexico
Merey del diablo	Venezuela
Bien me sabe; pan y quesito	Colombia
Aki	Costa Rica
Yeux de crabe; ris de veau	Martinique
Akie	Suriname
Kaka; fizan	Côte d'Ivoire
Finza	Sudan
Isin	Nigeria
Akye; akyen; ishin	Ghana; other African countries

tree have found various applications in traditional African and Jamaican medicine, such as the leaf extract in the treatment of ophthalmia and conjunctivitis, the ripe aril for dysentery, and the bark as a pain reliever (Gordon, 1999b). Ackee fruits are also used for fishing, for construction, and for several other purposes (Emanuel and Benkeblia, 2011). As discussed in Chapter 2, the aril is traditionally eaten in parts of West Africa, mainly Côte d'Ivoire and Burkina Faso, and in the Caribbean, mainly Jamaica, although its use has spread throughout the region and exports have now put the fruit on the menu in several North American, Canadian, and UK ethnic restaurants.

The fruit develops from blossoms (Figure 3.1b), which progress into small, hard green fruit that grow until they get to about 4–6 cm, at which time they become yellow or red in color (Figure 3.2a). These eventually ripen and open to expose the seeds and the arilli (arils). During this process, as the fruit is undergoing changes related to maturation, the hypoglycin toxin associated with the fruit also undergoes dramatic changes as described in the following section. In order to better characterize these changes and understand the stages of maturation of the fruit, Brown et al. (1992) developed a maturity scale for ackee that is still widely used throughout the industry and by regulators today (Table 3.2; illustrated in Figure 3.2). The understanding of the relationship between

Table 3.2. Maturity Scale Assigned to Ackee Fruit (*B. sapida*)

Assigned Number	Stage of Maturation	Description of Fruit	
1	Blossom developing	Tiny	Green fruit
2	Small	Hard green	
3	Medium	Hard green	
4	Large, full size	Some yellowing	
5	Fruit color has changed	Completely red or yellow (no green)	Red fruit
6	Slightly open	Pod lobes split to approx. 15 mm separation	
7	Medium open	Seeds and arilli visible	
8	Wide open	Pod interior visible	
9	Arilli completely exposed	Pod shriveling	
10	Onset of spoilage	Arillus decay evident	

Adapted from Brown et al., 1992, with permission.

the level of maturity of the fruit, its normal biochemistry including the changes that take place with hypoglycin A (HGA), and the impact that varietal, environmental, and other factors have on this are important to ensuring the safety of the fruit for consumption and trade, and can form a template for treating with other traditional products with natural constituents that require special care or handling.

3.2 HGA IN ACKEE: LOCATION IN THE FRUIT AND THE EFFECT OF MATURATION

Several studies have established the levels of HGA in different parts of the ackee (Bressler et al., 1969; Brown, 1989; Chase et al., 1990; Ellington, 1961b) and in ackees at various stages of the maturation (ripening) process (Brown et al., 1992; Chase et al., 1990; Hassall and Reyle, 1955; Manchester, 1974). The levels of HGA in the seeds, aril, and pods (husk) of the unripe (immature) fruit from Florida have been found to be 9390, 7110, and 416 ppm, respectively, decreasing to 2690 ppm in the seeds and less than 12 ppm in the aril, with the pods remaining unchanged (Chase et al., 1990). Levels of HGA in the immature aril ranging from 1000 to 6843 ppm have also been reported (Ashman, I., 1995, personal communications, BSJ; Brown, 1989; Hassall and Reyle, 1955), this depending on whether the immature fruit is at stage 3, 4, or 5 of the maturation cycle (Table 3.2, Figure 3.2). Stage 3 ackees typically having much higher levels of HGA than stage 5 ackees. Bowen-Forbes and Minott (2011) reported that HGA levels in the seeds and arils of "cheese" ackees (the harder variety) declined from 8000 ppm for stage 3 to 1451 and 271 ppm, respectively, for stage 6. Brown et al. (1992) found a rapid decline in HGA levels from just under 1400 ppm (stage 4) to under 30 ppm (stage 6). More recent studies by Gordon and Lindsay (2007) found a similar pattern, with HGA declining 10-fold from stage 4 to stage 6 (Figure 3.3). It is therefore well established that HGA levels decline in the fruit to very low levels as the fruit matures and becomes ripe.

Traditional practice, folklore, and even regulatory direction (Bureau of Standards Jamaica (BSJ), 2000) suggests that the raphe is a major source of the toxin (Stuart, 1975) and therefore needs to be removed carefully if the risk of illness is to be avoided. However, this has not been supported by research, with Brown et al. (1992) reporting that HGA levels in the raphe mirror those in the aril.

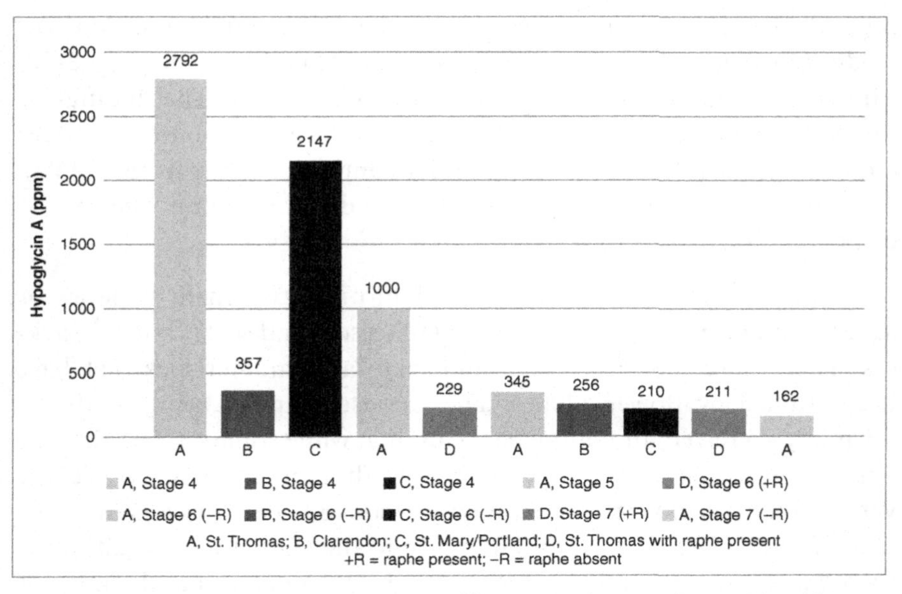

Fig. 3.3. HGA in ackees from four parishes in Jamaica at different stages of maturity. (Source: Adapted from Gordon and Lindsay, 2007, with permission.)

This finding has been reinforced by the data of Bowen-Forbes and Minott (2011), who reported HGA levels in the raphe at stages 5 and 6 that were similar to those in the aril. Gordon and Lindsay (2007), in a study of ackees from different locations across Jamaica, also found that for both stage 6 and stage 7 ackees, the presence or absence of the raphe did not significantly affect ($p > 0.05$) their HGA content (Figure 3.3, St. Thomas). These findings, and others (author's unpublished data), suggest that while the raphe, like the aril, contains high levels of HGA when the fruit is immature, its presence should not pose a food safety risk if derived from fully ripe fruit, beyond stage 7 in maturity.

3.3 IMPACT OF GEOGRAPHICAL LOCATION AND VARIETY ON HGA CONTENT IN ACKEE

There is a dearth of studies that have sought to characterize the effect of country of origin or geographical location on HGA content in ackee. Unpublished data on HGA in canned ackees from Haiti, Belize, and Jamaica (Technological Solutions Limited, 2012) indicated that country of origin did not significantly affect HGA content if the ackees were properly handled, in accordance with the FDA guidelines. Another study (Gordon and Lindsay, 2007) assessed the effect of the geographical origin of ackees across Jamaica on HGA content. The findings of this initial study also suggested that geographical location of the trees did not significantly influence the HGA content of the fruit ($p > 0.05$). Nevertheless, further work is needed to be definitive about the role of origin on HGA levels in ackees.

Traditionally in the Caribbean and Florida, two main varieties of ackees have been characterized, hard ("cheese") and soft ("butter") ackees. These have differing visual and textural characteristics that have been studied by Bowen (2006). The "cheese" variety, being harder, is normally preferred for commercial canning, while some suggest that the "butter" ackees have a better flavor, even though they tend to become very soft and of indifferent texture when cooked. The varieties from Belize, Côte d'Ivoire, and Haiti tend more to the cheese variety, although there has been no definitive work carried out to determine the ratio of "cheese" to "butter" ackee from these countries. Evaluation of the differences in HGA content between both types of ackee (Bowen-Forbes and Minott, 2011) revealed that immature "butter" ackees (stages 1–3) had a lower HGA content ($p < 0.05$) than "cheese" ackees. There was no difference, however, between mature "cheese" and "butter" ackees, with average HGA concentrations of 433 ppm in stages 5 and 6 for cheese ackees, compared to a mean of 481 ppm for "butter" ackees. As such, for ackees close to or at the stage at which they are prepared for consumption, variety appears to be unimportant to HGA content.

3.4 IMPACT OF GROWING SEASON AND REAPING AND HANDLING PRACTICES ON HGA LEVELS IN ACKEE

Technological Solutions Limited (TSL), a Jamaican private laboratory approved by the FDA and BSJ to do HGA analyses for exporters, reviewed data for canned ackees from Clarendon, St. Thomas, St. Catherine, and all other growing areas in Jamaica for multiple producers over the period 2002–2012 (unpublished data). These data showed that there was a tendency for the HGA content to increase in the first main season of the year, season 1, which runs from November to March and coincides with the dryer months, while being lower in the July to September crop, season 2. Bowen-Forbes and Minott (2011), on the other hand, found that HGA content was significantly higher ($p < 0.05$) for immature ackees in July to September (season 2) and lower in March (season 1), the opposite of what the TSL composite data showed. Differences may be due to varietal factors, or the fact that the TSL data were for canned ackees as against uncooked ackees for Bowen-Forbes and Minott (2011). Whatever the reason, further studies including the use of fruit from other countries are required to clarify the impact of season on HGA content.

While the seasonal impact on HGA content may not be definitive, it is more apparent that reaping and handling practices influence HGA content. Ackees from Haiti, Florida, and Belize are typically harvested directly from the trees, the fruit being picked when mature and open to stage 6 or 7. Côte d'Ivoire and Jamaica both harvest directly from trees and also take mature, unopened stage 5 ackees, which are placed on racks for ripening. In the case where mature, opened ackees are harvested directly from trees, the HGA content has been found to be typically below 100 ppm for stage 7 raw and cooked ackees (Brown et al., 1992; Chase et al., 1989). In some instances, however, stage 6 uncooked ackees were found to exceed the 100 ppm limit even though they were harvested directly from the trees (Bowen-Forbes and Minott, 2011; Gordon and Lindsay, 2007). Further, even fully mature (stage 7) ackees have been found to exceed the limit for both raw and canned ackees harvested in February 2006 (Gordon and Lindsay, 2007), a period during which the FDA also reported HGA levels to be high in canned ackees (Bliss, 2008; Martin-Wilkins, 2005). These ackees appear to have been affected by a general, island-wide increase in HGA levels for mature ackees, due to opened but immature ackees being used for processing (Figure 3.4a). This resulted from poor reaping, handling, and selection practices (Gordon and Lindsay, 2007). The details of this breakdown in practice are discussed further in Chapter 6.

3.5 THE ROLE OF THE SEEDS AND HYPOGLYCIN B IN DETOXIFICATION OF THE FRUIT

The levels of HGA in the aril of the ackee during maturation was studied in detail by Brown et al. (1992), who reported a rapid decline in HGA as the fruit moved from stage 5 to stage 6. Fowden (1975) suggested that HGA in the aril is translocated to the seed during ripening. Evidence

(a)

(b)

Fig. 3.4. (a) Old immature opened ackees; (b) stage 8 mature ackees with shriveling pods (husks). (Source: A. Gordon, 2015.)

that both aril and seed experience rapid falls in HGA levels at the same stages of maturity (stages 5–7, Bowen-Forbes and Minott, 2011; Brown et al., 1992) suggests that translocation alone is not responsible for the observations and that other factors are likely involved. Brown (1989) observed changes in the pH of both arils and seeds at different stages of maturity, with a sudden fall and rise in pH, respectively, for the arils and the seeds between stages 5 and 6, the stages at which HGA levels in the aril also fall rapidly and during which the pod begins to open. Speculation that HGA levels are related to fat content has not been verified, but what is clear is that the hypoglycin B (HGB) content of the seed rises as the HGA levels in both aril and seed decline during maturation. Bowen-Forbes and Minott (2011) demonstrated a highly significant inverse relationship between HGA content in the aril and the seeds of "cheese" ackees and HGB in the seeds during ripening (Pearson's r of -0.978 and -0.931, respectively). This relationship has received further evidentiary support by the findings that ackees with small or aborted seeds have significantly greater HGA content at maturity ($p < 0.05$) than those with regular seeds (Dundee and Minott, 2012), and that very small or aborted seeds (see Figure 3.1b) had significantly lower levels of HGB per gram than regular to medium-sized seeds. The HGB in the seeds therefore serves as a reservoir for HGA from the ripening arils and seeds and is thereby involved in the detoxification mechanism of the fruit (Bowen-Forbes and Minott, 2011). Detoxification is achieved by the translocation of HGA from the aril to the seed of the fruit and its subsequent conversion to HGB by γ-glutamyl transpeptidase (Fowden, 1975; Kean and Hare, 1980).

3.6 CONCLUSIONS

For fruits and vegetables that have been a part of traditional diets for many years and are now entering international commerce, but contain natural toxins, it is important to understand where in the fruit the toxin is located, what influences the level of the toxin, and the mechanisms of natural detoxification. Also important are factors which predispose the food to increased or reduced toxicity. All of these allow for science-based management of the handling and preparation of the food such that the risk of negative health consequences is eliminated or reduced to an acceptable level. For ackee (*B. sapida*), research has shown a systematic reduction in the toxin HGA to negligible levels when the fruit is fully mature (stage 7), and that this is accomplished by its translocation to the seeds. While geographical and varietal differences may also play a role, further study is needed to be definitive about this. What is clear is that much of the traditional knowledge has a sound scientific basis and because handling and reaping practices may influence the normal cycle of detoxification, these must be properly controlled if this fruit is to be commercially exploited safely and successfully. The same is also likely to be true of other horticultural products containing natural toxins.

CHAPTER 4

Biochemistry of Hypoglycin and Toxic Hypoglycemic Syndrome

André Gordon
Technological Solutions Limited, Kingston, Jamaica

ABSTRACT

This chapter presents an example of how detailed knowledge of the biochemistry of key components of concern in a traditional food containing a natural toxin, the ackee fruit (*Blighia sapida*), can facilitate better management of the treatment of any ill effects of its consumption, should they arise. The chapter therefore examines the biochemistry of the ackee fruit and its toxin hypoglycin A (HGA), as well as the nature of its toxicity. It discusses the illness associated with HGA, toxic hypoglycemic syndrome (THS), better known as Jamaican vomiting sickness (JVS), its etiology, symptoms, biochemical presentation, diagnosis, and treatment. It examines reports of illness in the Caribbean (Jamaica, Haiti, Suriname) and western Africa (Burkina Faso, Cote d'Ivoire), their authenticity, and the relationship between the types of ackee that have reportedly caused illnesses and HGA, and establishes the conditions under which toxicity is manifested in patients. The chapter also discusses how this knowledge can provide the basis on which regulators can agree on controls to ensure the safety of the food. The work that

Food Safety and Quality Systems in Developing Countries. http://dx.doi.org/10.1016/B978-0-12-801227-7.00004-4

has been done to document the effects of consumption of high levels of HGA and effective treatment of the resulting THS can form the basis of an approach that is also applicable to other traditional fruits and vegetables with natural toxins.

Keywords: ackee; toxin; biochemistry of hypoglycin A; Jamaican vomiting sickness (JVS); toxic hypoglycemic syndrome (THS)

4.1 INTRODUCTION

Many fruits and vegetables that are widely consumed contain natural toxins, antinutrients, or other toxicants that can be deleterious to health if not properly handled, as discussed in Chapter 2. Several of these are already being traded internationally, or form a part of diets around the world. Some that are an important part of tropical diets, such as cassava (*Manihot esculenta*), have become important in the growing trade between developed and developing countries. With the export of food and agroprocessed products now being seen as a major component of the sustainable economic development of many of the countries of the south (DaSilva and Baker, 2009; Regmi, 2001), the issue is not whether it is possible to trade in and handle these products safely, but the basis on which this can be done. For trade to be seamless, producers, exporters, food scientists, quality assurance and food safety specialists, and regulators must be able to agree on science-based approaches to the safe handling and trade in these products. It is therefore important that there is a shared understanding between the regulators of exporting and importing countries of the natural toxicants that are present in fruits and vegetables already being traded, or which have the potential for significant growth in exports, particularly where these are central to the economic well-being of the exporting country.

The previous chapter detailed how a toxin, hypoglycin A (HGA), in the ackee (*Blighia sapida*) is influenced by a variety of factors, which must be considered in controlling its level in the fruit. Chapter 5 will present the science-based approaches to eliminating or controlling HGA in ackees used for commercial production. The objective of this chapter is to provide an example, using the ackee fruit (*B. sapida*), of how detailed knowledge of the biochemistry of key components of concern in traditional foods can facilitate better management of the food in terms of the treatment of any ill effects of its consumption, should they arise. This can then provide the basis on which regulators can agree on controls to ensure the safety of the food. The ackee, the nature of its toxicity, the work that has been done to document and control this, and the issues surrounding the trade in the fruit can form the basis of an approach that can be used for other traditional fruits and vegetables with natural toxins. This chapter therefore examines the biochemistry of the ackee fruit, *B. sapida*, and the illness associated with HGA as a means of illustrating science-based approaches to the management of risk in handling traditional foods containing natural toxins.

4.2 BIOCHEMISTRY OF HYPOGLYCIN

Hypoglycin (HGA in earlier literature) is L-α-amino-β-2-methylenecyclopropyl propionic acid ($C_{77}H_{11}NO_2$) or, more simply, L-(methylenecyclopropyl)-alanine (Figure 4.1). Along with many useful nutrients such as the fatty acids linoleic, stearic, and palmitic; vitamins A and C, folic acid, and niacin; and potassium, sodium, calcium, and zinc (Lancashire, 2008; Odutuga et al., 1992), it is present in ackee. Hypoglycin (HGA), the major nonprotein amino acid in the fruit, is present in high concentrations in the aril and seed of the immature (unripe) fruit and differs from hypoglycin B (HGB), γ-glutamyl hypoglycin ($C_{12}H_{18}N_2O_5$),

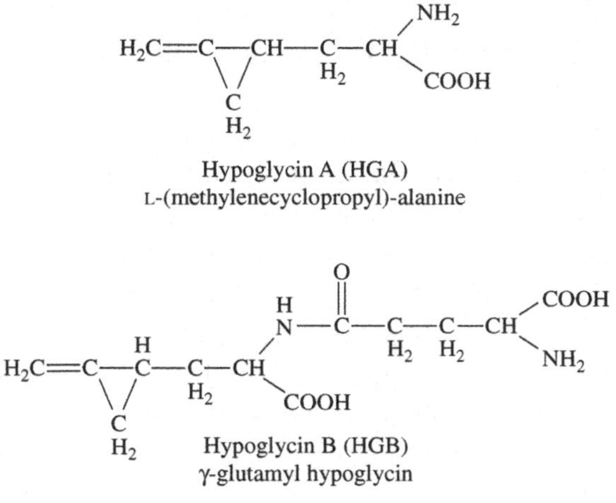

Hypoglycin A (HGA)
L-(methylenecyclopropyl)-alanine

Hypoglycin B (HGB)
γ-glutamyl hypoglycin

Fig. 4.1. Structures of HGA and HGB.

a dipeptide (Kean, 1989) that is typically found in high concentrations in the seeds (Bowen-Forbes and Minott, 2011; Dundee and Minott, 2012). Easier reference can be made to the compounds as HGA and HGB, respectively (Figure 4.1).

HGA is a neutral, atypical amino acid with both a cyclopropyl ring and a methylene group (Bressler et al., 1969; Manchester and Manchester, 1980). HGB is an acidic compound present in the seeds of the ackee (Ellington, 1961b; Fowden and Pratt, 1973; Kean and Hare, 1980; McGowan et al., 1989b). HGA is converted to HGB by γ-glutamyl transpeptidase (Kean and Hare, 1980), a reaction that occurs in the seeds (Bowen-Forbes and Minott, 2011). Both of these amino acids are similar in their chromatographic behavior to the amino acids leucine (Leu) and isoleucine (Ile) and, because of this, they were found to elute very closely to these amino acids during high-performance liquid chromatography (HPLC) analysis to determine the concentration of HGA (Sarwar and Botting, 1994; Ware, 2002). This may account for some of the challenges that earlier researchers encountered in quantifying HGA and hence some of the uncertainty about its presence in different parts of the fruit. This similarity to typical amino acids also influences the biochemical behavior of this atypical amino acid, particularly the way in which it is metabolized in the body. Metabolic processes that provide energy to the body can therefore use HGA as a precursor, and this is primarily responsible for its toxic nature.

Gluconeogenesis is the process by which the body produces glucose from noncarbohydrate precursors during periods of starvation or intense exercise when the body's normal stores and sources of glucose are not sufficient to meet its needs. This process occurs in the liver and the cortex of the kidney and is the principal means by which protein-precursor amino acids such as alanine are metabolized to produce glucose, and hence energy. Gluconeogenesis in the liver and kidneys helps to maintain the glucose level in the blood so that the brain and muscles can get enough glucose to meet their needs at all times. As a biochemical pathway, this process is an important alternative to oxidative phosphorylation (glycolysis; the process by which glucose, and hence the energy needed to sustain life, is usually produced), particularly during periods of starvation or severe malnutrition (Berg et al., 2012).

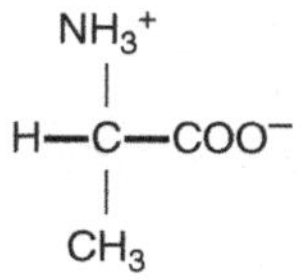

$$\mathrm{H{-}C(NH_3^+)(CH_3){-}COO^-}$$

Fig. 4.2. Structure of alanine.

Because of its similarity to the three-carbon amino acid alanine (Figure 4.2), HGA is also metabolized in the body through the gluconeogenic pathway typically taken by alanine. This pathway converts pyruvate to glucose, the pyruvate typically originating from lactate (from the muscles) or selected amino acids (including alanine). Pyruvate is converted to oxaloacetate by pyruvate carboxylase, which is located in the mitochondria of cells.

$$\underset{\text{Pyruvate}}{CO_2^- - CO - CH_3} + CO_2 + ATP + H_2O \xrightleftharpoons[\text{Carboxylase}]{\text{Pyruvate}} \underset{\text{Oxaloacetate}}{CO_2^- - CO - CH_2 - CO_2^-} + ADP + P + 2H^+$$

Oxaloacetate is an intermediate in both glycolysis and gluconeogenesis (Berg et al., 2012) and its concentration and availability in the mitochondria are therefore key signals to the body as to whether the glycolytic or gluconeogenic pathway is the main route required to produce glucose at a particular point in time. Pyruvate carboxylase is activated by acetyl CoA (or another closely related acyl CoA) and therefore depends on this intermediate being available for it to act. Anything that interferes with the availability of acetyl CoA or a related acyl CoA is therefore likely to have a deleterious impact on the levels of oxaloacetate in the cells and hence on the gluconeogenic pathway. HGA was identified as one such metabolite in a major review of the biochemistry of this unique amino acid by Kean (1989), with its primary mode of action being reaction with acyl CoA. This and subsequent studies led to further elucidation of the mechanisms by which HGA exerts its toxicity.

When it enters the body's digestive system, HGA (L-(methylene-cyclopropyl)-alanine) is metabolized by being transaminated to methylene cyclopropyl pyruvic acid (MCPP) (Billington et al., 1978), which is oxidatively decarboxylated to methylenecyclopropylacetyl coenzyme A (MCPA-CoA) (Figure 4.3). This requires and therefore reduces the availability of the α-ketoacid dehydrogenase complex that also assists in the degradation of valine (Val), Leu, Ile, and other branched-chain amino acids (BCAAs). HGA exerts its secondary effect on the gluconeogenic pathway by this means, impacting the blood serum and urinary concentrations of these amino acids (Lai et al., 1991) and reducing the availability of glucose from BCAAs. MCPA-CoA is also acylated by glycine *N*-acylase to produce MCPA-glycine (Figure 4.3), which is a major urinary HGA metabolite in rats (Tanaka et al., 1972; Tanaka and Ikeda, 1990) and also in humans (Kean, 1976; Tanaka, 1975; Von Holt et al., 1966). The final catabolites of HGA metabolism are MCPA-glycine and MCPA, both of which are normally excreted in the urine of ackee poisoning victims (Kean, 1989; Tanaka and Ikeda, 1990). Elevated levels of these compounds in the urine are

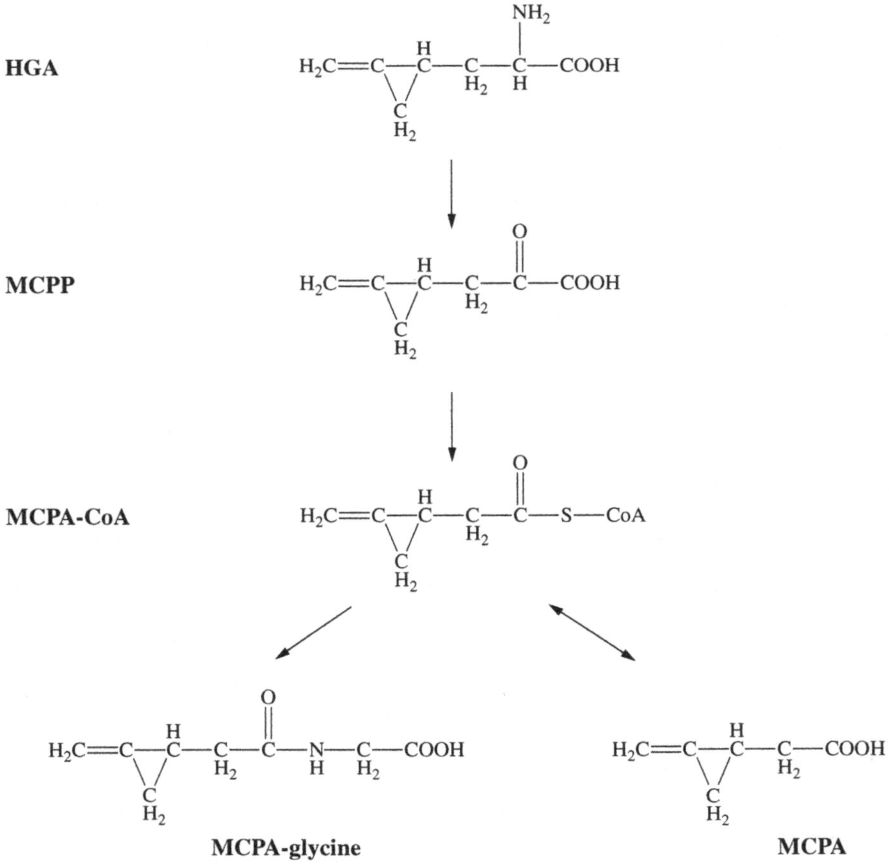

HGA

MCPP

MCPA-CoA

MCPA-glycine

MCPA

Fig. 4.3. Metabolism of HGA.

therefore typical indicators of the metabolic effects of HGA that occur during ackee poisoning.

The MCPA-CoA formed from the metabolism of HGA also impacts oxidative phosphorylation of short-chain fatty acids (SCFAs). It does this by irreversibly binding to flavin adenine dinucleotide (FAD), inhibiting the activity of the medium-chain and short-chain acyldehydrogenases, MCAD and SCAD respectively, with the 4R diastereoisomer being the more potent inhibitor (Lai et al., 1991, 1992, 1993). These enzymes are critical to the process of β-oxidation of fats and, therefore, inactivating them has adverse effects on blood serum concentrations of SCFA, as well as serum and urinary hydroxy and dicarboxylic acid concentrations (Lai et al., 1992, 1993). As such, there is a rise in the serum of free fatty acids (FFAs), such as propionic, crotonic, isobutyric, 2-methylbutyric, 3-methylcrotonic, and isovaleric acids (Tanaka et al., 1971, 1976). Decreasing β-oxidation of long-chain fatty acids (LCFAs) also decreases gluconeogenesis, resulting from a decrease in production of ATP, nicotinamide adenine dinucleotide (NADH), and acetyl-CoA. Elevated concentrations of isovaleryl-CoA, 2-14 methylbutyryl-CoA, glutaryl-CoA, and butyryl-CoA also competitively inhibits the activation of pyruvate carboxylase by acetyl-CoA and this is probably the major mechanism of inhibition of gluconeogenesis (Tanaka et al., 1972).

Accumulation of medium-chain dicarboxylic acids and hydroxylcarboxylic acids (C5–C10), such as 2-methylsuccinic, 3-hydroxyisovaleric, 2-ethylmalonic, glutaric, adipic, sebacic (C10), suberic (C8), hippuric, and stearic acids, and *n*-hexanoylglycine occurs in the blood serum and urine of both rats and humans intoxicated with HGA (Golden et al., 1984; Golden and Kean, 1984; Hine and Tanaka, 1984a,b; Shih and Tanaka, 1978). This is due to the excess LCFAs being supplied to the liver, which then undergo both mitochondrial and peroxisomal β-oxidation, producing an excessive amount of medium-chain fatty acids (MCFAs). ω-Oxidation then ensues via mixed-function oxidases produced by the endoplasmic reticulum, resulting in medium-to-short-chain dicarboxylic acids, which are subsequently excreted in the urine (Golden and Kean, 1984). Excretion of 3-hydroxyisovaleric and glutaric acids is a direct result of the inhibition of isovaleryl-CoA and glutaryl-CoA dehydrogenases, respectively (Hine and Tanaka, 1984a). Plasma and urinary changes in amino acids also occurred in rats treated with HGA. These include increases in plasma taurine, glutamine, glutamic acid, citrulline, tyrosine, ornithine, lysine, α-aminoadipic acid, and histidine concentrations in HGA-treated rats (Shih and Tanaka, 1978). It could be expected that similar observations will be made for people suffering from HGA intoxication.

Like other SCFAs, increased serum levels of isovaleric acid cause neurotoxicity (Rizzoli and Galzigna, 1970) and can produce symptoms of depression, vomiting, and ataxia (Tanaka et al., 1972). Increased levels of glutaric acid cause nephrotoxicity (Bressler et al., 1969; Harding and Nicholson, 1931) and may be responsible for the histological changes in animals' kidneys. Long-chain acyl-CoA esters are only oxidized as far as butyryl-CoA. The body compensates for a decrease in β-oxidation of fats by depleting liver glycogen and increasing the use of glucose for energy, resulting in hypoglycemia, this being the inescapable symptom of HGA poisoning.

4.3 THE ILLNESS ASSOCIATED WITH HYPOGLYCIN

The ingestion of immature (unripe) ackee arils or the drinking of the "pot water" (or brine) in which the ackee arils were cooked is known to cause an illness commonly called ackee poisoning in Jamaica (Golden et al., 1984; Kean, 1989). Because it was originally described in Jamaica, the illness is more widely known as Jamaican vomiting sickness (JVS). It is also, more accurately, called toxic hypoglycemic syndrome, THS (Baldwin and Parker, 1987), because HGA is the causative agent (Addae and Melville, 1988; Bressler et al., 1969; Ellington, 1961a) and there are reports of THS in which vomiting does not present (CDC, 1992). Confirmed cases of THS have also been reported in Haiti (Moya, 2006), Burkina Faso (Meda et al., 1999), Côte d'Ivoire (Foungbe et al., 1986), and Suriname (Gaillard et al., 2011). While immature ackee arils are known to cause THS, cooked mature (ripe) ackee fruit are substantially free of the toxin and are therefore safe to eat (Blake et al., 2004; Gordon, 1995a; Tanaka and Ikeda, 1990).

4.4 ETIOLOGY AND CLINICAL PRESENTATION

The conditions under which JVS occurs are well documented because of the long history of studying the illness (Ashurst, 1971; Stuart, 1975;

Tanaka, 1979). Hypoglycemia is an unvarying biochemical characteristic of the disorder (Stuart, 1975), hence the name THS. THS cases have a history of consumption of immature ackees within the previous 12 h. Typically in the reports from the Caribbean, all members of a family who share the meal become ill, regardless of sex or age. A universal feature of the illness is that malnourished individuals and children under the age of 15, especially if the latter are undernourished, are the most susceptible, with young children being disproportionate sufferers (Barennes et al., 2004; CDC, 1992; Foungbe et al., 1986; Kean, 1989; Kupfer and Idle, 1999; Moya, 2006; Tanaka, 1979). Meda et al. (1999) reported on an outbreak of epidemic fatal encephalopathy (EFE) in Burkina Faso, which they found to be caused by THS, resulting most likely from the consumption of unripe ackees. They identified 29 cases in five villages in the Toussiana and Péni divisions and neighboring areas of the country, all of children between the ages of 2 and 6 years, nearly all of whom exhibited symptoms of hypertonia (97%) and coma (100%), with 93% presenting with convulsions, 66% with vomiting, and only 7% presenting with fever. All died within 2–48 h and did not respond to treatment with quinine, antibiotics, or corticoids.

THS has been characterized by two main clinical types (Hill, 1952; Stuart, 1975). The first is characterized by the abrupt onset of vomiting 4–10 h after eating, without pain, fever, or diarrhea. There may, however, be some epigastric discomfort (Kean, 1989). Weakness and rapid prostration follows, after which there may be a period of remission. The vomiting or retching recurs, followed by drowsiness, twitching of limbs, convulsions, coma, and death, if the illness is not treated effectively. Dehydration is not normal. Recovery, if it occurs, is usually complete within 48–72 h, with no after effects. With the second clinical presentation, the illness progresses more rapidly and vomiting does not occur. The patient foams at the mouth, often with twitching of limbs, becomes drowsy, and collapses. Convulsions and fatal coma follow within a few days if not appropriately treated. The symptoms reported in the Burkina Faso and Côte d'Ivoire cases, while showing general similarities to the two main clinical presentations initially characterized, showed differences in presentation, and suggest that biochemical diagnosis is critical to establishing the cause of the symptoms. In all cases for which the data were available, severe hypoglycemia was found.

HGA and its metabolites interfere with gluconeogenesis, as shown earlier, resulting in a rapid depletion of hepatic glycogen, followed by severe hypoglycemia (von Holt et al., 1966). Further, the inhibition of isovaleryl-CoA dehydrogenase by HGA and its metabolites results in increased levels of branched-chain pentanoic acids, which are effective central nervous system (CNS) depressants (Tanaka et al., 1972). These authors suggest that the isovaleric acid (IVA) and α-methylbutyric acid (α-MBA) produced may be responsible for the vomiting and other CNS-related symptoms occurring in THS, even in the absence of profound hypoglycemia. More exhaustive discussions on the mechanism of HGA action are to be found in Kean (1989) and Tanaka (1972).

The laboratory evidence in confirmed THS cases invariably showed profound hypoglycemia, with blood sugar levels as low as 10 mg/100 mL (Stuart, 1975). Generally, serum glucose is reduced four- to fivefold (Kean, 1989). Acidosis often and ketosis sometimes accompany the other laboratory findings. Patients have been shown to have elevated serum levels of branched-chain pentanoic and other short-chain carboxylic acids, including IVA, α-MBA, glutaric acid, and ethylmalonic acid (Kean, 1989; Meda et al., 1999) and excrete MCPA and MCPA-glycine in their urine (Kean, 1989; Tanaka, 1979; Tanaka et al., 1976). Elevated serum levels of free fatty acids and the presence of amino acids (taurine, glutamine, tyrosine, and others) in the urine also occurs. Pathological findings include severe glycogen depletion and fatty metamorphosis of the liver (Stuart, 1975). These results, along with the clinical presentation and history of ackee consumption, are sufficient to confirm a diagnosis Jamaican vomiting sickness (THS).

Tanaka et al. (1972) suggested that ingestion of small amounts of HGA over long periods of time may result in chronic liver damage, especially since research has shown that the toxin remains in the bodies of animals for up to 24 h. They suggested that the occurrence of endemic, chronic liver diseases, such as veno-occlusive disease, gave credence to such a possibility. They also indicated that the etiology of these diseases is unknown. This theory was often repeated in the US literature regarding ackees in the 1960s and had, in part, formed the basis of the concern of the FDA, which led to the import alert on ackees (Gordon, 1999b). Liver biopsies of JVS/THS victims do show accumulation of fat in liver cells as an index of sublethal damage. This fat rapidly disappears as the toxin is cleared from the body, usually within a day or so, in response to glucose infusions (B. Hanchard, Department of Pathology, University Hospital of the West Indies, Kingston, Jamaica, personal communication received 1995). Patrick et al. (1955), using serial liver biopsies, had shown that liver glycogen is rapidly replaced following successful glucose therapy of THS victims. Further, Tanaka et al. (1972) observed that rats injected with HGA and treated with oral sucrose solutions showed no symptoms of IVA acidemia or hypoglycemia within 24 h. No elevated levels of MCPA or α-MBA were excreted or detected in the plasma in six out of eight rats. In addition, years of research at the Department of Pathology of the University Hospital of the West Indies in Jamaica has shown that the only form of dietary-related toxic liver disease is that due to the ingestion of herbal extracts, known colloquially as "bush teas" in Jamaica. These extracts are from *Crotalaria fulva* plants ("white back") and produce enlargement of the liver as a consequence of toxic damage by pyrrolizidine alkaloids on the lining of the blood vessels in the liver. The resultant obstruction to blood flow from the liver has occasioned the name "veno-occlusive disease" of the liver (B. Hanchard, personal communication). The clinical and pathological features of veno-occlusive disease have been well documented (Bras et al., 1954; Brooks et al., 1970). Consequently, while it is true that there were several chronic liver diseases of unknown etiology in Jamaica in the first half of the twentieth century, this is no longer true, nor is it true to suggest that chronic HGA consumption occasions such illness in Jamaica.

4.5 RECENT REPORTS OF TOXIC HYPOGLYCEMIC SYNDROME

The epidemiology of ackee poisoning has not been well characterized over the years and is only now getting more attention. In 1980, 271 cases were reported to the Jamaican Ministry of Health (JMOH) as a result of an outbreak (Chanel, 2001). Data on record indicate that, up to 1986, about 5000 deaths had been attributed to THS in Jamaica since the disease was first identified in the 1800s, the great majority of these occurring before the illness was fully understood. It is widely acknowledged that the frequency and scale of THS outbreaks in Jamaica have declined dramatically since the 1950s when early researchers were able to characterize the illness. Nevertheless, in late 1991 the JMOH requested assistance from the Centers for Disease Control (CDC) in Atlanta in the

United States, to investigate the incidence of the illness between 1989 and 1991 (CDC, 1992). The data gathered show that 28 cases, including 6 deaths, had been recorded as attributable to ackee poisoning. A high incidence was noted for male patients aged less than 15. The majority of the cases in Jamaica have historically been reported between January and March, although November to March, during which time ripe (mature) ackees are scarce, more accurately describes the period over which outbreaks tend to occur (CDC, 1992; Kean, 1989). The cases that have been reported in other countries, particularly in Côte d'Ivoire and Haiti, also occurred within a similar December-to-May period.

Cases in western Africa reported major symptoms similar to those reported in Jamaica. Over the December-to-March period, 1990–1997, Meda et al. (1999) reported 12 cases of ackee poisoning to have been confirmed in Toussiana, Burkina Faso. The largest outbreak of ackee poisoning reported in West Africa occurred in the same area of Burkina Faso between January and May 31, 1998, in which 29 cases of EFE were attributed to ackee consumption. The symptoms presented during this outbreak were characterized by vomiting, convulsions, or both, with progressive loss of consciousness and coma within 12 h, and with absence of fever at onset (Meda et al., 1999). These symptoms are similar to those typical of THS (Tanaka and Ikeda, 1990). All children stricken with the illness died within 2–48 h. Most patients were administered quinine ($n = 15$), antibiotics ($n = 11$), and corticoids ($n = 9$), none of which improved the course of the disease. In cases in which blood glucose concentration was measured, it was found that the patients were suffering from hypoglycemia. A survey conducted in the area of the epidemic showed that the ackee fruit was the common factor; furthermore, most people interviewed in the affected villages were unaware of the dangers of eating unripe (immature) ackee fruit (Meda et al., 1999).

According to Chanel (2001), ackee poisoning has been endemic in Haiti since 1988, a situation which recurred in Thibeau (Milot) in 1991. The problem again arose in late 2000 US FDA 2010a, when the CDC provided technical support to the Ministry of Health in Haiti during an outbreak of ackee poisoning in the northern region of that country (Joskow et al., 2006; Moya, 2006). More than 100 cases of acute illness and death were reported in Milot and Plaine-du-Nord, 80% of which were children (Chanel, 2001; Moya, 2006); the deaths resulted from eating immature ackees. In

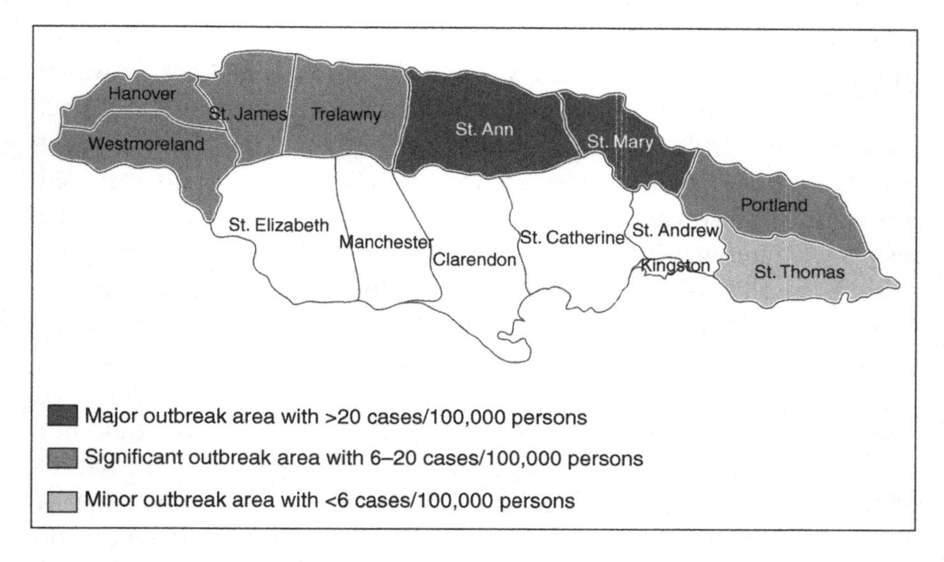

Fig. 4.4. Map of Jamaica showing major 2011 ackee poisoning outbreak areas. (Source: A. Gordon.)

Suriname, along the River Maroni that separates Suriname and French Guyana, reports detailed 16 deaths of children over the period from 1998 to 2001 linked to ackee fruit poisoning (Gaillard et al., 2011). This was as result of the misuse of the plant by Maroon witch doctors to "cure" some pathologies, especially acute forms of diarrhea in children. More recently, in December 2010 to May 2011, there was an outbreak in Jamaica reported by the JMOH in which there were at least 193 cases and 23 deaths (Wilson, 2011). Investigations suggest that the illness was due to the consumption of immature ackees (Gordon, 2011). The majority of the cases, which eventually exceeded 300, were localized along the northern coastal areas and concentrated in St. James, St. Mary, and St. Ann (Figure 4.4), which are not major ackee-growing areas. None of these Jamaican cases was linked to processed (canned or frozen) ackee fruit, the production of which is located in the southern part of the country.

4.6 SUSPECTED CASES IN EXPORT MARKETS

For many years, there were no reported cases of ackee poisoning from canned ackees. Indeed, none has ever been recorded in Jamaica or the UK, although canned ackees have been consumed in both countries for decades. In the United States, only two cases of possible ackee poisoning are on record. The first was in Toledo, Ohio, in 1994, in which a Jamaican woman presented with what was purportedly JVS after a meal of ackee fruit (McTague and Forney, 1994). The reported onset was 20 min after consumption of canned ackees. Symptoms included headache, profuse vomiting, weakness, and numbness. Laboratory test results were unremarkable except for elevated levels of adipic and lactic acids in the patient's urine. There was no evidence of hypoglycemia, nor were MCPA, IVA, or other common indicators of THS present. Many aspects of the case were largely atypical of HGA-induced vomiting, from the time of onset to the treatment administered. A stomach lavage was applied which resulted in almost immediate relief. This would not be effective in alleviating the symptoms of THS because of the nature of the illness (discussed earlier). It would, however, offer some relief to a patient suffering from other types of food poisoning. The diagnosis of JVS in this case is therefore curious.

The second case in the United States was in Connecticut, where a young Jamaican man presented with cholestatic jaundice, and his liver biopsy demonstrated centrilobular zonal necrosis and cholestasis, most consistent with a toxic reaction. The patient was interviewed regarding potential toxins, and he admitted to the ingestion of ackee fruit (Larson et al., 1994). The conclusion by the authors was that the symptoms were due to ackee consumption, although there was no indication of what other traditional foods he may have been routinely consuming, nor was the source or level of maturity of the ackee reported. Similar cases of liver damage due to the consumption of infusions ("bush tea") from *C. fulva* ("white back") had been well documented in Jamaica in the mid-twentieth century, as reported earlier (Bras et al., 1954; Brooks et al., 1970; B. Hanchard, personal communication).

To date, there has been only a single case of a possible illness associated with ackees in Canada (Gordon, 1999b). The illness was reportedly associated with canned ackees, although the level of HGA found in the cans was within that typically found in canned ackees (Brown, 1989; Chase et al., 1990), which have never been associated with ackee poisoning (Bliss, 2008; Gordon, 1999b). No other information on the case was provided, making it difficult to further evaluate the report. Consequently, while classic confirmed cases of ackee poisoning in major export markets from canned ackees are rare, vigilance continues to be required to retain the status quo.

4.7 TREATING ACKEE POISONING

Effective treatments for THS have been well established since the mechanism of toxicity was elucidated. Glycine, riboflavin, intravenous glucose, and clofibrate have been used to help alleviate the adverse effects of the toxin HGA (Al-Bassam and Sherratt, 1981; Kean, 1974; Sherratt, 1969). Intravenous glucose restores a low blood glucose level to normal; however, it does not always prevent death (Kean, 1974), depending on time of administration after the onset of illness. Administration of riboflavin, on the other hand, helps to replace the chemically modified flavin prosthetic groups of inhibited enzymes, thus assisting the body to replace the modified coenzyme FAD (Sherratt, 1969). Administration of glycine facilitates conjugation of the glycine moiety to MCPA, thus forming the excretory product MCPA-glycine (Al-Bassam and Sherratt, 1981). This limits the toxic effects of hypoglycin and minimizes SCAD and MCAD enzyme inhibition by lowering cellular concentrations of MCPA-CoA. Finally, when clofibrate was incorporated into the diets of rats treated with hypoglycin, it caused an increased number of peroxisomes and an increase in peroxisomal β-oxidation, thus reducing the need for the body to resort to blood glucose for energy (Kean, 1974; van Hoof, 1985). This treatment may also be useful for human patients suffering from THS. In summary, administering glucose, glycine, and riboflavin, singly or in combination, shortly after the onset of THS can ameliorate the symptoms and reverse the negative effects of the illness. Clofibrate will also help. These treatments for THS are well known in Jamaica and if they are used in combination with proactive and aggressive public education campaigns during the period of November to May, issues with THS should be minimized. The deployment of these approaches in other countries, which have had outbreaks should also have similar success and significantly reduce the incidence of fatalities.

CHAPTER 5

Effective Science-Based Approaches to Establishing Safe Food Handling Practices for Traditional Foods: The Ackee Example

André Gordon
Technological Solutions Limited, Kingston, Jamaica

ABSTRACT

This chapter aims to present a summary of what we now know about the ackee, bringing together traditional and anecdotal knowledge with hard scientific data, with a focus on commercial aspects of the ackee, production methods, and risks, both perceived and real. Detail is given of the handling and processing of ackees, the regulations and standards that govern these processes, and the science behind them. Hazard analysis critical control points are discussed. The information presented here could serve as a template for similar approaches with other traditional foods containing natural toxins being exported to developed countries.

Food Safety and Quality Systems in Developing Countries. http://dx.doi.org/10.1016/B978-0-12-801227-7.00005-6

Keywords: ackee; critical control points (CCPs); food safety; hazard analysis critical control points (HACCPs); HGA; quality assurance; raw material handling and processing; regulatory oversight; safe-handling practices

5.1 INTRODUCTION: BACKGROUND TO THE SAFE PRODUCTION AND HANDLING OF ACKEE (*Blighia sapida*)

In this book so far, there has been an examination of the ackee fruit (*Blighia sapida*), including environmental and other factors, the impact of specific peculiarities of the fruit, and also detailed discussions on the biochemistry and toxicity of the fruit. The science behind the fruit has been explored in Chapters 2, 3, and 4, and it should be recognized that while it contains a natural toxin, hypoglycin A (HGA), it can still be consumed like other fruits and vegetables as it is benign when it is fully mature (Brown, 1989). It has already been noted in Chapter 3 that several factors can influence the content of the toxin in the fruit. Manipulating or controlling these will make the fruit safe for consumption. Chapters 6 and 7 will delineate how the science behind the fruit that has already been discussed, as well as new findings and approaches, was used to lift the import alert imposed on ackees by the United States Food and Drug Administration (US FDA). Nevertheless, prior to much of the work discussed in this book being carried out, most of the reports in the literature on ackees had been made on the hypoglycins HGA and hypoglycin B – HGB, mainly on HGA, the toxin associated with Jamaican vomiting sickness (JVS), and its biochemistry and the mechanism by which it exerts its toxic effects (Ellington, 1961a; Manchester, 1974; Stuart, 1975; Tanaka and Ikeda, 1990). Other, later, studies examined the levels of the toxin in processed and fresh ackees (Brown et al., 1992; Chase et al., 1990). However, much of the information available on commercially processed ackees has never been published, having been gathered in the commercial sphere and by regulators, and used for problem solving rather than as a tool to enlighten scientists, regulatory practitioners, horticulturalists, exporters, and others about how such information can be used to build an export industry for a traditional food containing a natural hazard.

The objective of this chapter, therefore, is to bring together all of the information available on this unique fruit to present a comprehensive overview of the scientific foundation for the industry that has developed around *B. sapida*. It will bring together the decades-old traditional and anecdotal knowledge, old wives' tales, and hearsay, which have been studied in an effort to verify their veracity, and, where appropriate, synthesize these with the hard scientific data that have significantly expanded understanding of how this traditional fruit has behaved commercially, particularly since 2000. The nature of the toxin, hypoglycin, the illness associated with it, and the various mechanisms involved in its metabolism and by which it exerts its effects have already been discussed in detail in Chapter 4. Consequently, these will not be a major focus of this review, having been already adequately addressed, here and in other reviews (Kean 1988, 1989; Sherratt, 1986; Tanaka and Ikeda, 1990). The focus here will be on the commercial aspects of the product, the scientific basis for the method of production used, the levels of toxin found in commercially processed ackees, the real and perceived risk of injury to health resulting from consumption of the fruit, and the approaches for risk minimization, such as the hazard analysis critical control points (HACCP)-based programs used in the industry. It is hoped that the information presented here and its application to the commercial processing of this traditional fruit will catalyze similar approaches to other traditional foods destined for developed-country markets.

5.2 THE MARKET, CONSUMERS, AND EXPORTS

Ackees have been a part of traditional western African and Jamaican cuisine many years, with the consumption of the fruit dating back to the nineteenth century (Ashurst, 1971; Kean, 1989). As has been noted previously, the fruit is also eaten by people throughout the Caribbean, in North America, and the United Kingdom, largely in areas populated by Jamaicans or by people with Jamaican connections, but has also been exported to many other countries (Gordon, 1999b). In recent years, there has been significantly greater interest in the fruit for its culinary uses; because of its notoriety; because it is a standard breakfast menu item at Jamaican hotels, exposing it to millions of visitors; and also because of its growing economic significance in Haiti, Belize, Côte d'Ivoire, and Jamaica, as well as commercial entities in other countries known to export occasionally or who may be contemplating exports. These countries include Costa Rica,[1] the Dominican Republic, and Ghana.[2] The fruit has, therefore, been consumed over the years by millions and, like potatoes, aubergine, or cassava, and as discussed in Chapters 2 and 4, there have been relatively very few instances of adverse consequences.

Jamaica began exporting ackees to Canada and the United Kingdom in 1957, and has been exporting to various countries around the world ever since. Exports have also gone to Japan, the Caribbean, Holland, Central and South America, and Guantanamo Bay and, up until 1993, frozen ackees went to the United States until they were also included under the import alert, which has subsequently been lifted. Today, the major export markets remain the United States, the United Kingdom, and Canada. Exports of canned ackees from Jamaica over the first three decades were between 40,000 and 80,000 cases per annum, with the amount increasing substantially to exceed 140,000 cases per annum in recent years, the United States having once more become a destination country (Bliss, 2008). Ackees are distributed mainly in urban areas in the United States, Canada, and the United Kingdom, and are carried by several major retail chains as well as the more traditional ethnic food stores. Major buyers include Publix in the United States, Tesco and Sainsbury's in the United Kingdom, and Metro and Loblaws in Canada. The main consumers of the product remain Jamaicans and other people of Caribbean descent. However, because of the wide exposure of tourists to the product during their visits to the island and ackees being regarded as a Jamaican delicacy, the product has begun to attract mainstream consumers in these markets.

He who eats the isin (ackee) should know how to remove the poison.

A West African proverb

As is evident from this proverb and some of the names in Table 3.1 (Chapter 3), the methodology for the preparation and safe consumption of ackees is very well known to those for whom it is a part of their diet.

---

[1]There have been persistent reports about occasional illegal exports to the United States from Costa Rica, where the fruit also grows, and the FDA have been approached by potential exporters there as to how to gain access to the US market (Saltsman J., 2006, personal communication FDA, Maryland, USA; Gordon A. 2004, personal communication, Technological Solutions Limited, Kingston, Jamaica).

[2]Since 2005, commercial interests from these countries have approached the author's business, Technological Solutions Limited (TSL), for discussions and assistance in setting up commercial operations.

This is because the preparation and consumption of ackees are part of the native customs of Jamaicans and those from the parts of Burkina Faso and Côte d'Ivoire where the fruit is eaten. Therefore, it is widely accepted that properly handled, cooked arils from mature (ripe) ackees present no health hazard (Brown et al., 1992; Tanaka and Ikeda, 1990), whereas consumption of immature ackees is dangerous, especially to malnourished children (Bressler, 1976; Centers for Disease Control (CDC), 1992; Stuart, 1975). It has further been shown that for commercially processed ackees that have ripened naturally on the trees or opened on ackee-ripening racks (Figure 5.1a), the fruit is completely safe (Bliss, 2008; Brown et al., 1992; Chase et al., 1990; Gordon, 1999a; Ware, 2002).

Traditionally, ackees are prepared for home consumption by first selecting fit, ripe fruit that have opened naturally on the tree. These ackees would be classified as stage 7 or stage 8 (Table 3.2, Figure 5.2a,b), using the classification system developed by Brown (1989). Immature (unripe) ackees (stages 2–4) and mature (ripe) ackees (stage 5) are not used. The ripe ackee arils (Figure 5.1b) are removed from the pods, cleaned by removing the seed and the pink raphe (also called the "membrane"), and immersed in boiling water. The water is then decanted and the parboiled ackee arils are flavored with pepper, salt, and other spices and cooked, usually by frying in oil, with meat or fish. This may be, for example, salted mackerel, canned corned beef, or, in the case of Jamaica's national dish – *ackee and saltfish* – salted codfish. Commercially, this process is replicated by using brine as the heating medium and cooking in the can (i.e., thermally processing the ackees), and ackees canned in brine are by far the major variant for producers in Côte d'Ivoire, Haiti, Jamaica, and Belize. Some freezing of ackees is practiced in Jamaica and Florida, and both individually quick frozen (IQF) ackees from Jamaica and frozen ackees from Florida have been offered for sale in traditional markets, including in the US market since 2000 (Gordon, 1999a).

5.3 RAW MATERIAL HANDLING AND PROCESSING

Ackee was originally not an orchard crop, being harvested in the wild in western Africa and Jamaica, although there were some areas where there were natural large stands of trees that could approximate an orchard. There were also, from the 1990s, a few people in Jamaica who sought to

Fig. 5.1. (a) Ackee ripening rack; (b) ackee arils (with seed attached). (Source: A. Gordon, 2015.)

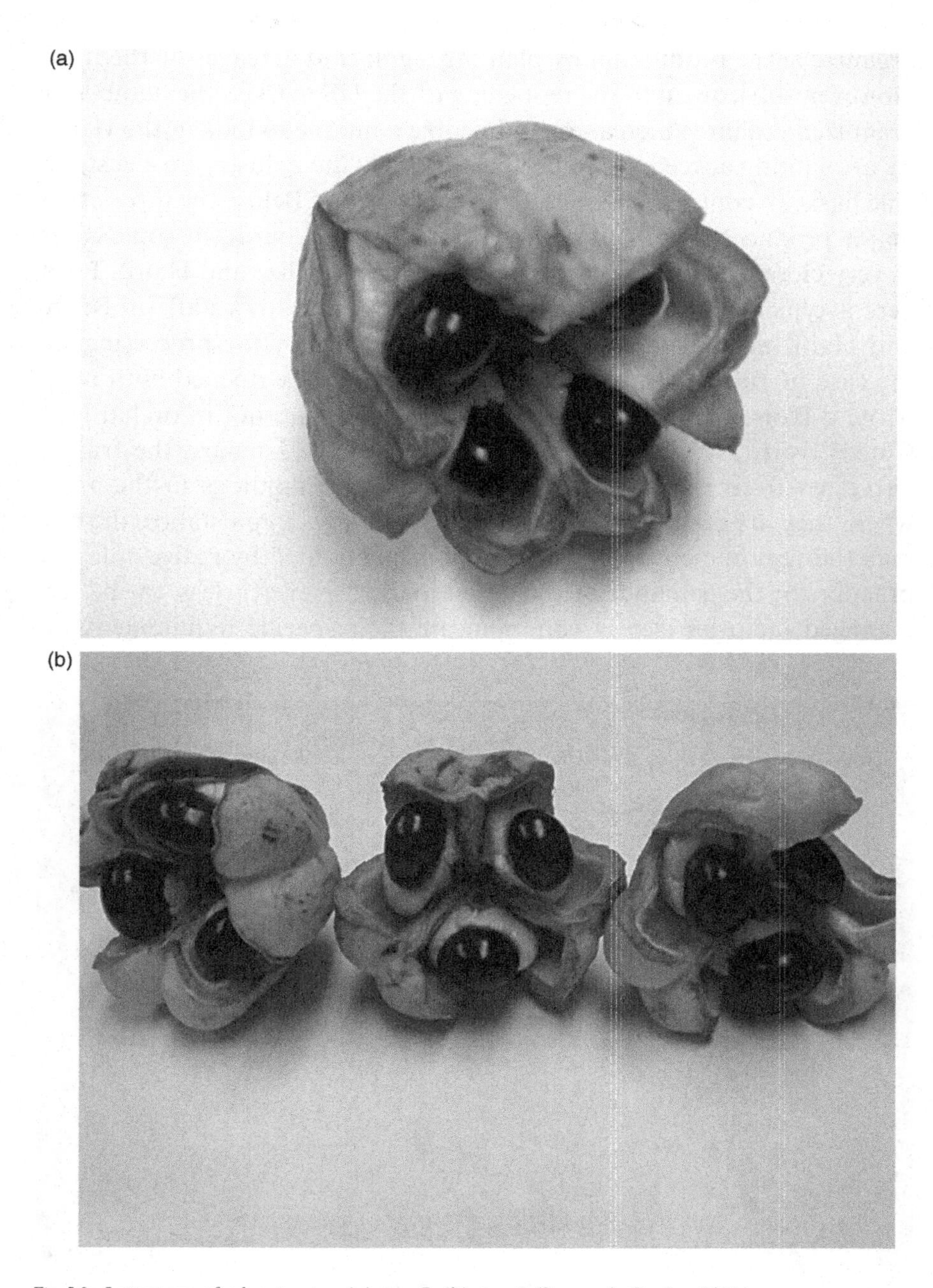

Fig. 5.2. Later stages of ackee ripening: (a) stage 7; (b) stage 8. (Source: A. Gordon, 2015.)

organize ackee production by planting significant acreages of the fruit. However, subsequent to the reopening of the US market, the number of organized orchards increased, and people continue to look at the viability of planting ackee trees to be able to supply the industry on a sustainable basis. In contrast, in Haiti, Côte d'Ivoire, and Belize, the three other major producers, the fruit is grown mainly in orchards, in some cases in very close proximity to the processing plant (Belize and Haiti). Fruit here is typically harvested by the processing company's staff (in Belize and Haiti) and taken either directly to the factory for processing (in the case of ripe (mature) open fruits) or to be rack ripened until open (Côte d'Ivoire). In those cases where the ackees are not in orchards in Côte d'Ivoire, or where there are no orchards in Jamaica, the fruit is harvested from trees on private or public land holdings in the areas where they are concentrated (Figure 5.3). In any circumstance, the raw material requirements are well known, being guided by native culinary practices or the demands of the target market. Nevertheless, the better-managed factories clearly communicate their specific requirements to

Fig. 5.3. Harvesting ackees from trees. (Source: A. Gordon, 2015.)

raw material buyers and maintain records of the quality of the fruit being provided on a supplier-by-supplier basis.

In Jamaica where the source is often itinerant suppliers, opened, mature, and fit fruits (stages 5–9, Table 3.2, Chapter 3) are the ones chosen for processing. Fit, mature fruits are typified by full, firm pods (stage 5) or fully opened fruit (stages 7 and 8, Figure 5.2) up to stage 9 of maturity (Brown, 1989). A typical, generalized process for ackees canned in brine,[3] similar to what was routinely used during the period 1998–2002 for processing ackees prior to the approval of the FDA, is shown in Figure 5.4. This was the flow chart initially submitted to the FDA (Gordon, 1999b). Selected steps in the process are also illustrated in Figures 5.5, 5.6, 5.7, and 5.8, while Figure 5.9 shows a process that is similar to what is currently being used in Jamaica.

Ackees are picked (Figure 5.3), collected by contracted vehicles at different harvesting sites, and transported to processing facilities. The period between harvesting and delivery to the factories typically does not exceed 24 h. On delivery, each load is inspected by experienced staff to determine whether the fruit maturity requirements, as characterized by Brown et al. (1992) and codified by the Bureau of Standards Jamaica (BSJ)[4] in Jamaican Standard (JS) 276 (BSJ, 2000) and an illustrative chart issued by them, have been met. At the time, this was done by cutting open and examining the raphe and arils of a preset number of unopened ackees from each load to determine whether they were fit and would open within the specified period of 3 days (stage 5, Figure 3.2a, Chapter 3). Loads deemed "immature" were rejected, being of no use to the processors since they would not ripen and could not be processed. There was therefore a very strong commercial incentive to suppliers to take in ackees that were either fit or had begun to open on the trees. This inspection at receival was designated an important critical control point (CCP), CCP 1 (Figure 5.4) in the Hazard Analysis Critical Control

[3]The process is generalized as there are substantial differences in the way the canned fruit is produced in Haiti, Belize, and Côte d'Ivoire, where the process involves acidification and aril hardening, prior to processing with a much more benign, less aggressive, thermal process.

[4]The Bureau of Standards Jamaica (formerly the Jamaica Bureau of Standards) is the regulatory body responsible for the processed food industry in Jamaica under the Standards Act (1969) and the Processed Foods Act (1959).

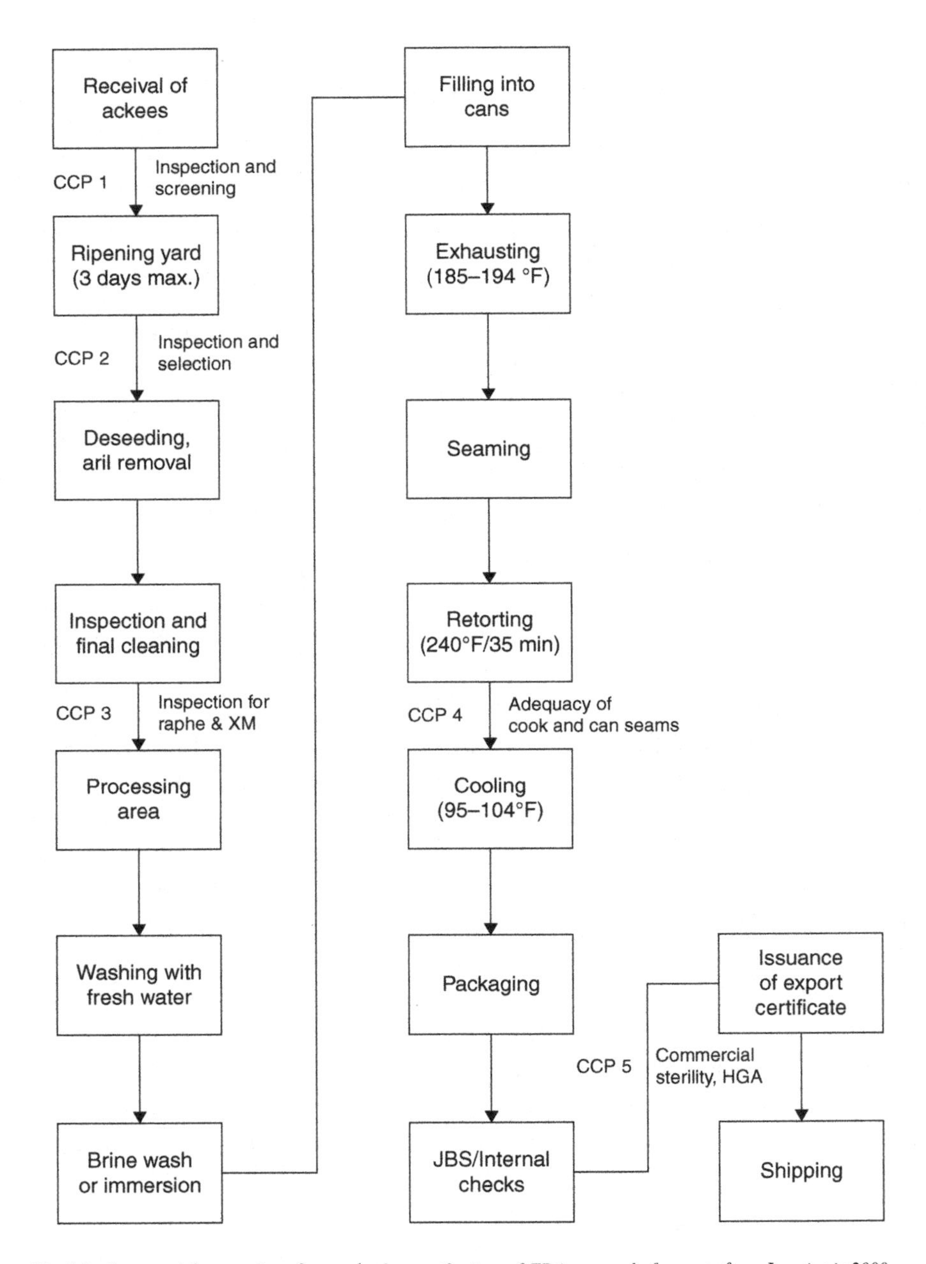

Fig. 5.4. Commercial processing of canned ackees at the time of FDA approval of exports from Jamaica in 2000. (XM, extraneous matter.) (Source: Adapted from Gordon, 1999b, with permission.)

Fig. 5.5. Steps in ackee processing: (a) racking; (b) selection from the racks. (Source: A. Gordon, 2015.)

Fig. 5.6. Steps in ackee processing: (a) shelling (husk/pod removal); (b) final inspection prior to canning.
(Source: A. Gordon, 2015.)

Fig. 5.7. Steps in ackee processing: (a) filling cans; (b) adding brine. (Source: A. Gordon, 2015.)

(a)

(b)

Fig. 5.8. Steps in ackee processing: (a) exhausting; (b) seaming cans; (c) retorting; (d) cooling. (Source: A. Gordon, 2015.)

Fig. 5.8. (Continued)

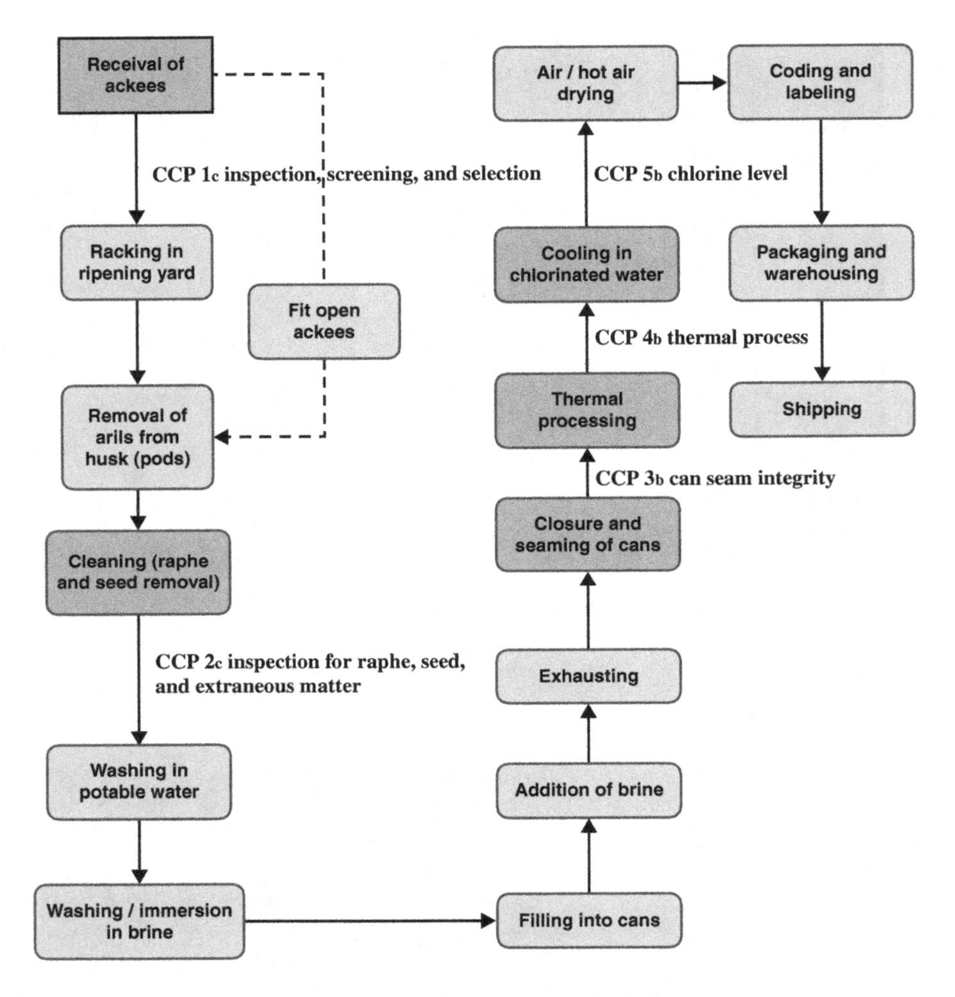

Fig. 5.9. Outline of current commercial processing of canned ackees (generalized).

Points (HACCP) plan for Ackees Canned in Brine in 1999, as it was critical for control of the hazard HGA, which could get into the finished product if not substantially controlled at this initial step of the process (Gordon, 1999a,b).

Once a load was accepted, it was sorted and unopened fruit placed on tiered racks in the sun to ripen (Figure 5.5a). Open fruits were sent directly to the shelling (husk (pod) removal) operation (Figure 5.6a), if mature (fit), and then on to cleaning. In some operations, cleaning and shelling were done in the same area. Processors typically noted the date of receipt of the ackees and tagged the racks accordingly to facilitate traceability, a requirement instituted in 1999. Traditional practice and research have both shown that fit (stage 5) fruit will ripen and open on racks exposed to sunlight within 3 days (Bates, 1991; Brown et al., 1992). During the 3-day period allowed for ripening, opened ackees were constantly removed to the sorting and cleaning operation by people assigned to monitor the fruits on the racks (Figure 5.5b). These people were responsible for management of the rack-ripening process, rotating incoming fruit onto the racks and opened fruit or fruit that had not ripened off, and were specially trained and experienced in identifying ackees at different stages of maturity (Gordon, 1999a), particularly since this was a CCP (CCP 2, Figure 5.4). The racks were cleared 3 days after initial racking of the ackees, on an ongoing basis (Figure 5.4).

In the cleaning area, the arils were removed from the pods, the seeds and the raphe (membrane) were removed, the arils were inspected (CCP 3), and the cleaned arils transferred to the processing area. The arils were then washed, including a brine-immersion stage, packed into cans (Figure 5.7a), which were then filled with brine (Figure 5.7b), exhausted[5] (Figure 5.8a), and sealed (CCP 4, Figures 5.4 and 5.8b), prior to a typical retorting procedure[6] (Figure 5.8c), which concluded with the cooling of cans in chlorinated water (CCP 5, Figure 5.8d).

The scheduled process used by the industry varied, but a typical process as determined by temperature distribution studies and heat penetration studies (determination of the slowest heating point of the canned product) was 35 min at 240°F, with an F_o[7] of 6. This process was one of the options presented by the recognized process authority,

---

[5]Exhausting is the process by which steam at about 170–180°F is used to drive air out of the tissues of fruits and vegetables prior to closure of the cans/glass containers in which they are to be processed. It helps to shrink the fruit or vegetable and increases the vacuum formed when the product shrinks on cooling.

[6]Retorting is the process in which products are heated and cooked using steam under pressure to achieve a much more rapid and effective cook capable of destroying spores of mesophilic and thermophillic organisms of public health or economic significance.

[7]This is a measure of the lethality of the process and indicates its ability to destroy the most heat-resistant forms of the microorganisms of public health and economic significance being targeted – *Clostridium botulinum* and other spores.

Tech-S Corporation from the United States, working under the auspices of the Food Technology Institute (FTI) and Bureau of Standards in Jamaica (BSJ) in the early 1980s. Subsequent processes used in the industry included 11 min at 250°F, with an F_0 of 5.2, developed by Technological Solutions Limited (TSL), and others developed by the FTI (Gordon, 1999a). All were theoretically sound, based on the principles of Stumbo (1973) and Pflug (1987), and fully met the requirements of 21 CFR 113, the FDA low acid canned food (LACF) regulations.

Subsequent to Jamaican ackees regaining access to US markets, the good manufacturing practices (GMPs) and HACCP-based food safety program and the process used, including some of the CCPs, have changed for some processors (Figure 5.9). This is because some processors have refined their process and now undertake all activities where the arils are handled under more hygienic conditions inside the plant, with their second and final control for HGA, CCP 2, at the cleaning stage (Figure 5.9). They have also placed greater focus on controlling incoming fruit and have combined this with selection from the racks to make one CCP (CCP 1, Figure 5.9), which they indicate has made the original CCP at the selection step (CCP 2, Figure 5.4) redundant as a CCP. This has been validated as being effective for those plants that have it in place. Others continue to use a HACCP plan very similar to the original one shown in Figure 5.4, which is also validated as being effective.

5.4 THE SCIENCE BEHIND THE INDUSTRY: THE IMPORTANCE OF RESEARCH, REGULATORY OVERSIGHT, AND SCIENTIFIC AND TECHNOLOGICAL SUPPORT FOR THE EXPORT OF TRADITIONAL FOODS

5.4.1 Food Safety and Quality Assurance: Regulatory Oversight

The process of manufacturing canned ackees in Jamaica is regulated under the Processed Food Act, 1959 (revised in 1964), the Jamaican Standard Specification for Processed Food (General), JS 36: 1991, and the Jamaican Standard Specification for Canned Ackee in Brine, JS 276 (2000), all compulsory standards under the Standards Act (1969). Regulatory control is based on the available evidence that if ackees are handled and processed in the manner prescribed, they present no health hazard to consumers (Bates, 1991; Brown, 1989). The Jamaica Bureau of

Table 5.1. Regulatory Requirements for Export of Canned Ackee from Jamaica Until 2000[1]

Chemical/Physical	Can Evaluation	Microbiological
Color (>75% uniformity) Flavor/odor Net weight, drained weight Defects (as % not properly cleaned) Character (texture) % Unit wholeness (>75%) pH, vacuum, headspace	External can defects Legibility of coding Seam integrity Tightness rating	Commercial sterility at 35°C (95°F)[2] Commercial sterility at 55°C (131°F)

[1]Companies approved to export to the United States underwent assessment of the LACF and food safety programs and were exempt from these requirements, except for the new requirement of testing for HGA.
[2]As prescribed by the *Bacteriological Analytical Manual* (BAM) of the FDA.
Source: Gordon, 1999b.

Standards (JBS), working with the FTI, had set standards for the ripening and processing of canned ackees under the Processed Food (Grades and Standards) Regulations (1964), including the variables to be monitored. A subsequent standard, JS 36, requires that critical control points should be monitored. These CCPs were, however, not specified, nor had the existence of a functioning, verified, and validated HACCP program been made a specific requirement of the processing and export of ackees under this standard. This is an area that was later addressed in 2000 with the promulgation of JS 276, largely to govern exports to the United States. Export of ackees to other destination countries, however, still fell under the pre-existing standards and required all of the tests indicated in Table 5.1 to be performed with satisfactory results, prior to the approval of the product for export.

Every lot of ackees that was legally exported from Jamaica was, therefore, required to be tested by the BSJ (the Processed Food Act, 1964) or other approved laboratories[8] up until 2004 when some companies became exempt from batch-by-batch testing based on adherence to a verified and validated HACCP program. For all exporters, regardless of the standard under which they operated, the BSJ issued export certificates if the product met the required standards. The Food Inspectorate Department also visited and inspected ackee-processing facilities, including the can seam records, process charts and records,

[8]These are Technological Solutions Limited (TSL) and the Mona Institute of Applied Sciences (MIAS).

and other CCP and water chlorination records as part of their monitoring practices. This is still in place today. However, as Jamaica sought to further update its food regulatory system to accord with the new imperatives of global trade, the BSJ issued a new standard in 2012, the Jamaican Standard for the Production of Processed Foods – Utilizing the HACCP Principles (General), JS 317:2012. This standard strengthened the requirements under JS 276 and extended them to all foods.

In addition to the above requirements and the tests listed in Table 5.1 required to be performed by nonexempt processors, all LACF processors are required to ensure that their process is filed with the FDA and the BSJ and is under the supervision of someone who has successfully completed a BSJ-approved thermal processing course of instruction. This course is currently the FDA-approved Better Process Control School (BPCS), hosted in Jamaica by the University of the West Indies. There are also ongoing training programs for ackee processors, supported by industry associations. In Belize, the processing company is regulated, as are all other foods, by the Belize Agricultural Health Authority (BAHA) and must meet basic food safety requirements, although not as extensive as those for Jamaican exporters. In other jurisdictions, such as Haiti and Côte d'Ivoire, none of this kind of regulatory oversight exists. The FDA set up a system for monitoring Haitian exports before they were allowed to enter the US market, but neither the Canadians nor the United Kingdom has specific monitoring programs for ackee imports entering their markets.

5.4.2 Food Safety and Quality Assurance: The Science Behind Industry Practices

The ackee fruit, its characteristics, and the commercial aspects of its handling and processing have, over the years, been the subject of significant scientific studies that have helped to address some of the major concerns, including those that led to the banning of the product from the United States. Henry (1994) discussed these and also undertook a risk assessment of the ackee fruit for the FDA (Henry et al., 1998). At that time, the toxin associated with the fruit and the illness caused by it had been well characterized, but there was not enough information about the science supporting commercial practices in the production of ackees. Henry (2006) did update the assessment of the previous reports, but a subsequent synthesis of the studies that have provided scientific support for the commercial practices in the ackee industry is not in the public domain. This section therefore presents the status of the scientific background to industry practices and examines other aspects of the commercial practices that have not yet been covered.

The HGA content of the parts of the ackee has been studied by several researchers over the years, with Hassall and Reyle (1955), Ellington (1961a,b), and Manchester (1974) being among those reporting on the content in different parts of the fruit. The HGA levels in ackees as they mature and ripen have also been well documented (Bowen-Forbes and Minott, 2011; Brown et al., 1992; Chase et al., 1990; Gordon, 1999b; Gordon and Lindsay, 2007; Gordon et al., 2006). It is therefore well accepted that the HGA content declines such that once ackees reach stage 6, it is usually well below the level of concern, 100 ppm (Figure 5.10). There are circumstances, however, where the HGA level appears to rise above the norm, resulting in HGA levels at stage 6 in excess of 100 ppm (Chapter 3). These include the seasonal rise in HGA levels previously reported (J. Kerr, BSJ, 1999, personal communication; Bowen-Forbes and Minott, 2011; Technological Solutions Limited (TSL), 2014, unpublished data), as well as the rise in HGA levels in the fruit due to poor harvesting practices and abuse of handling resulting in immature, but open, ackees (Figure 5.11) becoming comingled with mature, normally opened ackees (Campbell, 2006; Gordon and Lindsay, 2007).

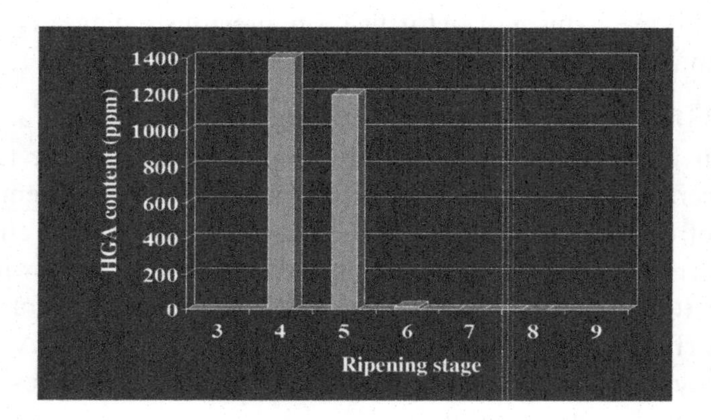

*Fig. 5.10. HGA content of ackee (*B. sapida*) at various stages of maturity (ripeness).*

Fig. 5.11. Unfit (immature) ackees – various stages. (Source: A. Gordon, 2015.)

As discussed previously, Bowen-Forbes and Minott (2011) and Dundee and Minott (2012), respectively, reported on the effect of the seasons, variety, and seed size on the HGA content of the fruit, while the presence of the raphe was explored by Brown (1989), Bowen-Forbes and Minott (2011), and Gordon and Lindsay (2007). These latter studies suggest that the traditional belief that the raphe may be an area of high HGA content is not supported by the evidence, but more work is needed to validate these findings. While earlier studies did not consistently support the widely held belief that arils with atypical (in color) or aborted seeds were particularly dangerous (Bates, 1991), other subsequent studies have shown this to be the case (Dundee and Minott, 2012). Bowen-Forbes and Minott (2011) examined the role of the seed in acting as a reservoir receiving HGA from the aril in the maturing fruit as it is converted to HGB by transamination, and provided strong evidence to support this mechanism of detoxification, previously suggested in other reports, including Fowden (1975) and Kean (1989).

The BSJ, FTI, and University of Florida had undertaken a series of studies in 1988–91 to help to understand the science of the fruit and its commercial processing, out of which came initial refinements of the method of quantification of HGA (Chase et al., 1989, 1990), characterization of the fruit and the seminal maturity index chart (Brown, 1989), and a better understanding of the scientific impact of commercial practices (Bates, 1991). The study also investigated the HGA levels in different types of ackee arils and arils with different histories. The results of these investigations were HGA levels ranging from less than 0.1 to 897 ppm, depending on the nature of the samples and the stage of

Table 5.2. Selected Early Studies Reporting the Levels of Hypoglycin in Canned Ackees

Hypoglycin in Canned Ackees (mg/100 g)		
Arils	Brine	References
4.3–23.2	6.15–34.35	Sarwar and Botting (1994) (Health and Welfare, Canada)
19.89–20.83	26.04–27.18	F. Jamieson (1995, personal communication) (Health Protection Branch, Health and Welfare, Canada)
0.74–9.08	0.8–11.1	I. Ashman (1991, personal communication) (BSJ/USAID Project)
<1.2	–	Chase et al. (1990)
0.2–11.4		J. Kerr (BSJ, 1999, personal communication)

maturity of the fruit. The HGA content of canned ackees, which was also evaluated under this project, had also been the subject of several studies; the results of some of these are summarized in Table 5.2. Values ranging from 0.13 to 208.3 ppm in the arils and 0.35–326 ppm in the brine had been reported at the time. These early studies on the commercial aspects of the fruit and the industry provided the basis for further, more focused, studies on aspects of the industry that were scientifically undocumented.

Data from the BSJ for ackees produced in the 1994 winter season and scheduled for export show levels ranging from 2.86 to 12.84 ppm for the aril and 2.94–21.08 ppm for the brine (Gordon, 1999a). Data for 1994–1996 indicate mean levels in the range of 11.21–61.39 ppm in the arils (Table 5.3), with the overall mean value being approximately 27.20 ppm (J. Kerr, BSJ, 1999, personal communication). Data for subsequent years

Table 5.3. Hypoglycin Levels in Arils of Canned Ackees for Export, 1994–1996,[a] 1999,[b] 2005,[b] 2010[b]

Year	No. of Samples	Mean HGA Level (ppm)
1990	48	33.59
1994	39	11.21
1995	39	17.82
1996	20	61.39
1999	28	58.26
2005	62	72.53
2010	98	56.91

Sources: (a) J. Kerr, BSJ, 1999, personal communication; (b) Gordon, 2014, personal communication.

show a rise in mean HGA levels, but these remained below the regulatory target of 100 ppm.

Research into ackee processing revealed that HGA shows great thermal stability, requiring 131 min at 250°F (D_{250}[9] = 131) to obtain a 10-fold reduction in concentration (Brown, 1989). Such a process would be so deleterious to product texture as to render the product unpalatable. Consequently, once HGA is present in the arils, its removal from the fruit by cooking is not a viable option. Brown (1989) also investigated the changes in HGA levels at different stages of the canning process with a view to determining what effects the process had on HGA. It was widely held that, because of the solubility of HGA, it may be possible to remove appreciable amounts of the toxin during the washing stages of processing. Brown (1989) found that while leaching of the toxin did occur (0.4–2.14 ppm), it was not significant enough to be of major importance. The immersion time required to significantly impact HGA levels, in excess of 2 h, was too great to be practical, especially considering the susceptibility of ackees to softening if held immersed for extended periods of time. However, Golden (2006) reported that soaking and cooking in brine reduced the HGA content of arils, an observation which had been made previously in studies done by the BSJ (Kerr, 1995; 1999, personal communication) and which has also been supported by a subsequent study, although the effect of soaking was not specifically assessed (Gordon and Lindsay, 2007).

In studies on the leaching of HGA from arils to the brine in canned ackees, the levels in the brine exceeded the HGA content of the arils (Bates, 1991; Sarwar and Botting, 1994). Other studies, however, found the HGA levels in the brine to be between 31 and 45% of the total in the can (Chase et al., 1989; Scott et al., 1974). Overall, the existing canning procedure does not appear to further reduce HGA levels in ackee arils (Brown, 1989). It is possible that the differences reported could have resulted from the time between manufacture and equilibration of the HGA content between the two phases. Because the data are not conclusive,

[9]Decimal reduction time at 250°F – time required at 250°F to kill 90% of the microorganisms being targeted. In this case, these were bacterial spores of *Clostridium sporogenes*, PA 7639.

more research is required on this. Further studies would therefore need to be done to determine the effects of processing on the reduction of the HGA content of the arils and whether other factors can improve the efficacy of the process in reducing HGA levels from the fresh to the processed product.

The microbiology of the fruit, in general, and the potential impact of the microorganisms present on the efficacy of the thermal process applied to canned ackees has not received much attention. Nevertheless, Gordon and Jackson (2002) investigated the microbiological flora of the fruit during ripening and found that microbial numbers were, in general, low, except at stage 10 where the fruit was rapidly deteriorating. At this stage, *Staphylococcus* spp., lactic acid bacteria, and yeasts and molds were the dominant microorganisms. Another finding was that spore loads were typically low. Gordon and Jackson (2013) later reported that for rack- and tree-ripened ackees, the microbial content and flora found suggested that rack ripening could, in some cases, result in better quality fruit for processing, as a result of a more controlled environment that reduced the incidence of and opportunities for contamination of the fruit. More importantly, the very low mesophilic spore loads found on both tree- and rack-ripened ackees at all stages of maturity supported the use of a much more benign thermal process (F_0 of 3.0) than is typically targeted (F_0 of 6.0), as in the Tech S and FTI processes mentioned earlier. This finding lends credence to the widely held assertion in the Jamaican industry, supported by decades of anecdotal evidence, that even thermal processes previously regarded as marginal by the authorities will achieve sufficient lethality to render the products commercially sterile. This merits further investigation because of its implications for commercial practices and food safety.

5.5 QUANTIFICATION OF HYPOGLYCIN A (HGA)

Hypoglycin has been a problem to extract and quantify in a reliable manner (Kean, 1989; Manchester, 1974). In fact, the difficulty in accurately quantifying the levels of HGA present in ackee had been one of the major concerns that the FDA had with allowing its importation into the United States (Bates, 1991; McGowan et al., 1989b). Most of the methods in use prior to the late 1980s were based on paper or thin-layer chromatography (Ellington, 1961b; Kean, 1989) and had several limitations. Scott et al. (1974) used ion exchange amino acid analysis (IE-AAA) to quantify HGA in canned ackees. However, this method has as a major limitation in its inability to resolve HGA from other amino acids, such as leucine (Leu) and isoleucine (Ile), with similar solubility and chromatographic properties.

Chase et al. (1988, 1989) developed and used an improved IE-AAA method that gave better resolution of amino acids to quantify the HGA levels present in canned ackees. This method used postcolumn ninhydrin derivatization, a cationic resin, and has a detection limit in the range of 4.8 ppm. McGowan et al. (1989a), working with the BSJ, developed a reverse-phase high-performance liquid chromatography (RP-HPLC) amino acid analysis technique with precolumn derivatization with O-phthalaldehyde (OPA). This method was more effective and versatile than the IE-AAA methods and had a detection limit of about 0.1 ppm, as well as good interlaboratory reproducibility. Sarwar and Botting (1994) adapted an HPLC method, the pico-tag method, to determine the HGA levels in canned ackees imported into Canada. The method, which had earlier been developed for rapid amino acid analysis (Sarwar et al., 1988), employed a precolumn derivatization with phenylisothiocyanate (PITC) and gave very good resolution of HGA, Ile, Leu, and other amino acids. It had the advantage of shorter elution times than the McGowan et al. (1989b)/BSJ method (6 min vs. 30–35 min). The detection limit with this method is 10 ng. Ware (2002) validated the Sarwar and Botting (1994) method, and Whitaker et al. (2007) refined the sampling plan to develop the one currently being used by the FDA and all approved laboratories for testing canned ackee in brine.

5.6 SAFETY OF ACKEE AS A FOOD

The issue of the safety of ackee (*B. sapida*) as a food has been at the root of the concerns and intrigue with this fruit since being raised by Tanaka et al. (1972). The many years of commercial operation of the industry and regulatory oversight have provided copious anecdotal support for the safety of the fruit, as there has never been a confirmed case associated with canned ackee (Gordon, 1999a,b). Several studies (including those discussed above) have also shown that ackees canned as is currently practiced in exporting countries are safe because the HGA levels are well below the regulatory limit when mature fruit are used (Bates, 1991; Blake et al., 2004; Brown et al., 1992; Chase et al., 1990). In fact, in a study of the fruit and its consumption, specialists concluded that the canned product presented a minimal food safety risk (BIBRA, 1995). They indicated that "it is unlikely that even the ingestion of a whole can (arils plus the liquid), which would provide a maximal dose for an adult of about 2 mg/kg BW, would have overt toxic effects." This is also supported by the findings of a study carried out by the University of Florida (Bates, 1991).

Henry et al. (1998) addressed the concerns of the FDA with the information available at the time, and the risk assessment was sufficiently favorable for the FDA to consider the issue and eventually lift the import alert on canned ackee. And this was before information from the studies on ackee and HGA consumption by Jamaicans (Blake et al., 2004) and on the maximal tolerated dose (MTD) for rats (Blake et al., 2006) became available. These data showed that the MTD, if approximated to humans, would be more than 150 times the daily intake reported for the highest consumers of ackee. Substantial abuse of the handling of the fruit and consumption of excessively high levels of immature fruit in the canned product would therefore be required for there to be a realistic risk of acute or chronic toxicity from HGA poisoning from canned ackee.

CHAPTER 6

Re-entering the US Market with Jamaican Ackees: A Case Study

André Gordon[1], Joyce Saltsman[2], George Ware[3], and James Kerr[4]

[1]Technological Solutions Limited, Kingston, Jamaica
[2]Retired from the Center for Food Safety and Applied Nutrition (CFSAN), US Food and Drug Administration, Baltimore, Maryland, USA
[3]Retired from the Southeast Regional Laboratory (SRL), US Food and Drug Administration, Atlanta, USA
[4]Retired from the Bureau of Standards Jamaica, Kingston, Jamaica

ABSTRACT

The ackee has great socioeconomic importance to Jamaica, providing vital employment opportunities, so the import alert (ban) on its importation into the United States in the 1970s due to safety concerns was a significant blow. This chapter discusses the efforts made by Jamaica and the United States to overcome the market access challenge, which resulted in the lifting of the import ban in 2000. The activities and approaches initiated by the Jamaica Exporters Association (JEA)-led public/private partnership and the Jamaican Ackee Task Force to meet the requirements of the US Food and Drug Administration (FDA) for removing the import alert it had imposed on ackees, are detailed. Also detailed are the approaches undertaken by the FDA and the scientific effort required to support the process in both countries. This process, as well as

Food Safety and Quality Systems in Developing Countries. http://dx.doi.org/10.1016/B978-0-12-801227-7.00006-8

leading to the removal of the import alert, had many other benefits to both Jamaica and the United States, and can serve as an example for future approaches to such trade challenges.

Keywords: Bureau of Standards Jamaica (BSJ); Center for Food Safety and Applied Nutrition (CFSAN); Good Manufacturing Practices; hazard analysis critical control points (HACCP); import alert; Jamaica Exporters' Association (JEA); Jamaican Ackee Task Force (JATF); low-acid canned food (LACF) regulations; US Food and Drug Administration (FDA); US Department of Agriculture (USDA); ackee

6.1 INTRODUCTION

Ackee, Jamaica's national fruit, is not only a staple of the Jamaican diet, but is also socioeconomically important as it is grown and processed in rural areas where employment opportunities are limited. The ackee-processing industry is therefore very important to Jamaica. *Ackee and codfish*, Jamaica's national dish, was ranked number two in the world by a National Geographic survey of national dishes (National Geographic, 2011). The fame and international renown of the fruit is matched by an intriguing history, fraught with challenges that can provide a template for dealing with difficult market-access issues for traditional exports to global markets. As has been discussed in previous chapters, ackee poisoning resulting from the ingestion of unripe ackee arils or from drinking the water in which the arils were cooked causes the illness known as Jamaican vomiting sickness (JVS) or toxic hypoglycemic syndrome (THS). This concern about the possible presence of the toxin hypoglycin A (HGA) in the fruit prohibited its importation into the United States from 1973 (Brown, 1989; Gordon, 1999b; US Food and Drug Administration (US FDA), 2000), although Jamaica was exporting to various other countries around the world such as Japan, Central and South America, the United Kingdom, and Canada. Despite this, the product was available in the United States, but was imported illegally into areas where there were high populations of people of Jamaican descent, such as Florida and New York.

In 1998, the Jamaica Exporters' Association (JEA) was approached by the US embassy in Jamaica to develop and implement a program to get Jamaican ackees back into that market. This came out of a close working relationship with the US Department of Agriculture (USDA) at the time, which translated into a personal interest of the ambassador, Stanley McLelland, hoping to make a transformational impact on the Jamaican economy. The JEA formed what was called the Jamaican Ackee Task Force, led by Director Dr. André Gordon, which spearheaded the process (The Daily Gleaner, 2000). Some months later, in April 1999, at the request of the US embassy in Jamaica, the FDA met with the US ambassador, the US embassy agricultural attaché, representatives of the Bureau of Standards Jamaica (BSJ), the JEA, and industry representatives to discuss the food safety issues relative to the fruit. The participants presented information on safe handling and processing practices to reduce the levels of toxin, and outlined a program for their processors, with oversight by BSJ, to ensure that preventive controls were in place.

This was the start of an alliance between the US competent authority (the FDA), the Jamaican competent authority (the BSJ),[1] the ackee industry, and the exporters' trade association (the JEA), which began as a simple outreach program to develop a preventive controls program to assure the safety of imported ackee into the United States (Saltsman and DeVlieger, 2012). Today, the FDA enjoys a robust working relationship with both competent authorities in Jamaica, the BSJ and the Ministry of Agriculture, as well as the ackee industry and the JEA, and the goal of ensuring safe ackee products from Jamaica has been achieved. This chapter explores and documents the processes and considerations involved in addressing this significant bilateral trade challenge as a case study that can advise future approaches to trade challenges.

6.2 BACKGROUND TO THE MARKET-ACCESS CHALLENGE: THE IMPORT ALERT

The ackee and ackee and saltfish (codfish) have been consumed by Jamaicans and millions of visitors to the island for years, as previously noted. As has also been noted, the issues related to the safe consumption of the fruit, the toxin HGA, and the associated clinical condition JVS

[1]The Bureau of Standards Jamaica (BSJ), known as the Jamaica Bureau of Standards (JBS) at the time, is empowered under the Standards Act (1969) and the Processed Food Act (1959) to regulate the production, sale, and export of *prescribed foods*, which must meet preset requirements to be eligible for export.

or THS, have been known since the nineteenth century. Public records between 1885 and 1887 indicate that deaths attributed to ackee poisoning were noted by the island chemist (Hill, 1952). The handling of the fruit caught the attention of research scientists including Hassall and Reyle (1955), who isolated HGA and hypoglycin B (HGB) from it. Ellington (1961a,b), Kean (1974), and Tanaka (1979) reported the range of concentrations for HGA in both immature (unripe) and mature (ripe) ackee arils and, subsequently, it was established that if the use of immature fruit was avoided, there was very little risk of illness (Brown, 1989). Because Jamaicans knew about the fruit and its dangers, the handling of ackees for local consumption and, subsequently, for export as a canned product, was always done in a manner that minimized the risk. In fact, ackees had been exported from Jamaica to the United States and other countries since the 1950s (J. Kerr, 2014, BSJ, personal communication) and, as noted in Chapter 4, there has never been a confirmed case of THS linked to legally imported canned ackees in any of the importing countries.

Despite this history of safe consumption for commercially handled and exported ackees, Tanaka et al. (1972), in assessing the potential chronic toxicity of HGA, suggested that there was a "likelihood that small amounts of Hypoglycin A ingested over time may cause cellular injury and possibly liver damage." The authors made a connection between some chronic liver diseases endemic to Jamaica and unspecified veno-occlusive conditions and the ingestion of HGA. They concluded that HGA may be responsible for the pathogenesis of the observed liver diseases, even though earlier work in the 1950s and 1960s had already established that these diseases were due to plant toxins linked to the consumption of *Crotalaria fulva* in form of traditional teas (Bras et al., 1954). The Tanaka et al. (1972) publication raised alarm bells at the FDA as they considered HGA to be a food additive for which safety data had not yet been established (i.e., it could not be classified as "generally recognized as safe" – GRAS). Once aware of the fruit's importation into the United States and the potential public health risk associated with hypoglycin, the FDA regarded ackee as an adulterated product that would require scientific proof of its safety; therefore, the agency imposed an import alert on ackees in 1973 (Bliss, 2008). The alert was upgraded to "ackees in all forms" in 1993 (US FDA, 2000).

The BSJ was the competent authority in Jamaica responsible for the regulation of the food industry, and was tasked with the responsibility to address the FDA's concerns. At the time, the FDA indicated four conditions that would provide science-based assurances that the FDA felt were sufficient to allow it to amend the import alert and permit firms to import the product into the United States (J. Kerr, 2014, BSJ, personal communication; Saltsman and DeVlieger, 2012). They were:

1. The development of a reliable and reproducible method of detecting and quantifying HGA and HGB
2. Data indicating that safe levels (or virtually none) of the toxic amino acids were present in fruit meant for human consumption
3. Assurances that only ripe fruit would be used for processing
4. If residual levels of HGA and HGB were present in fruit to be consumed, the levels would not be harmful (i.e., present only at a "no-effect level").

The BSJ then initiated several activities and approaches geared toward meeting the requirements stipulated by the FDA, so that ackee could once again enter the US market (Table 6.1). These included undertaking the development of the method for detecting and quantifying

Table 6.1. Key Milestones in the Successful Lifting of the Import Alert on Ackees

Year	Action	Outcome
1973	FDA Import Alert 21-11 imposed on canned ackees (*Blighia sapida*) because of insufficient evidence of its freedom from HGA and HGB; HGA classified as a non-GRAS food additive	Legal exports of canned ackees and ackee products to the United States prohibited; Firms automatically on "detention without physical examination" (DWPE), which means any product arriving at a US port was automatically denied entry without any inspection of the product and had to be shipped elsewhere or destroyed
1988	Commencement of the Jamaica Ackee Project (JAP) involving preparation of a verifiably pure HGA standard, validation of the analytical method, and studies of the commercial product	Successful extraction and purification of HGA by the JBS Chemistry Department based on a modification of a method developed by Kean (1974)
1989	Work on a validated method for the determination of HGA concluded at the University of Florida (UF) under the JAP	Publication of the method in McGowan et al. (1989a)
1990	Chemical studies on commercially canned ackees undertaken to properly characterize the product	Ongoing application of the HPLC-based methodology developed by the UF team for HGA First set of studies on commercially canned ackees completed and information compiled for the JAP final report

(continued)

Table 6.1. Key Milestones in the Successful Lifting of the Import Alert on Ackees (*Cont.*)

Year	Action	Outcome
1991	Studies on ackees, including canned ackees and ackee ripening, completed at UF	1. Establishment/formalization of a maturity index for ackees 2. Documentation of the levels of HGA associated with each stage of maturity 3. Establishment of likely levels of HGA in canned ackees 4. Publication of milestone research by the UF team in Brown et al. (1992) 5. Database of HGA results on canned ackee from across the island established
1992	Ongoing monitoring of HGA levels in canned ackees commenced in earnest	Database of HGA results on canned ackee substantially expanded by new data
1993	Visit to the FDA by Jamaican delegation to review the import alert[1]	FDA interested in evaluating analytical method by IUPAC and indicate more detailed evidence required for the safety of HGA as a food additive
October 1998	Jamaica Ackee Task Force (JATF),[2] under the auspices of the JEA and supported by the US embassy through the USDA, is established	A comprehensive program to transform the industry to make selected processors HACCP compliant and HACCP-based regulatory oversight is developed
April 1999	Jamaican delegation led by the JEA visits the FDA and presents a comprehensive science-based petition for the lifting of the import alert[3]	The FDA agrees to a program which, if successful, will allow the removal of the import alert for qualified exporters
April/May 1999	FDA sets up Jamaica Ackee Outreach Program[4]	Key collaborations are initiated; objectives and goals are set for Jamaica and the FDA to work collaboratively for the first time
October/ November 1999	The first FDA inspection team visits Jamaica to inspect plants and visit the JBS	A relationship between the industry and regulators in Jamaica and the FDA, supported by the US embassy, is established
July 6, 2000	The import alert on ackees canned in brine is lifted	Canco Limited and Ashman Food Processors Limited are able to commence exports to the United States

HACCP, hazard analysis critical control points; HPLC, high-performance liquid chromatography; IUPAC, International Union of Pure and Applied Chemistry; JBS, Jamaica Bureau of Standards; USAID, United States Agency for International Development.

[1]The delegation consisted of the late Dr. Juliette Newell of Tijule Ltd. (representing the processors), Dr. George Wilson from the Jamaica Agricultural Development Foundation, and Mr. James Kerr from the JBS, and was supported by the US embassy, the Jamaican embassy, and USAID.

[2]The JATF was chaired by Dr. André Gordon, Managing Director of Technological Solutions Limited, a JEA director, and included representatives of the Ministries of Industry and of Commerce and Agriculture, the JBS, processors and distributors, funding agencies, and the USDA. The detailed composition of the task force is given in Appendix II.

[3]This delegation included the Jamaican ambassador, Dr. Richard Bernal, the USDA, the JEA, processors, and the BSJ. The technical and industry presentation was led by Dr. Gordon and included Mr. James Kerr (JBS) and Mr. Norman McDonald (Canco Ltd.).

[4]The Jamaica Ackee Outreach Program was led by Dr. Joyce Saltsman, Interdisciplinary Scientist, Office of Plants, Dairy Foods, and Beverages (now the Office of Food Safety).

HGA and HGB, other data gathering, continued tight regulation of the industry, and a mission to the FDA in 1993. However, these were not successful in providing the kind of science-based assurances that the FDA felt were sufficient to allow it to lift the import alert. It was not until 1998, with the coming into being of a focused national effort and close collaboration with the FDA, that an approach that had the agreement of all parties was crafted.

6.3 AGREEING AN APPROACH TO ADDRESSING THE IMPORT ALERT

The BSJ developed and implemented a series of studies and programs aimed at meeting the FDA's requirements for lifting the import alert on this critical export item from Jamaica. These complemented private-sector-driven activities. The series of programs and approaches undertaken by the BSJ, the JEA, and their collaborators provided much of the commercial information, data, and regularization of the industry required to support a successful thrust towards re-entry of ackees into the US market. These, in addition to the steps taken in 1999 (summarized in Table 6.1), formed the basis of a good case study for collaboration on market-access issues for traditional products between countries. The full sequence of actions that led to the lifting of the alert are presented in Appendix I.

A JEA/US embassy initiative led to a private sector/public sector delegation to the FDA on April 19, 1999. The delegation consisted of two JEA directors, JEA members, a representative of the processors and a major distributor in Canada and the UK, the Jamaican-based USDA representative, a representative of the Jamaica Bureau of Standards (JBS), and the Jamaican ambassador to the United States. The meeting with the FDA, held at the agency's Center for Food Safety and Applied Nutrition (CFSAN) in Washington, DC, on April 20, 1999, was chaired by the US ambassador to Jamaica, His Excellency the Honorable Stanley McLelland. At the meeting, a detailed review of commercial processing practices and the science behind the fruit as known at the time (see Chapter 5 for a summary of this) was presented. Data presented on hypoglycin levels of Jamaican canned ackees, which had been collected by the JBS since 1991, represented a critical part of the presentation. The current and proposed regulatory oversight, the latter now being based on hazard analysis critical control points (HACCP)-based production, and the willingness of the industry, the Government of Jamaica (GOJ), and their collaborators, the USDA, to do whatever was necessary to meet the FDA's requirements, were emphasized. The FDA assembled a team of scientists including toxicologists; food safety specialists; plant scientists; and specialists in low-acid canned food (LACF) production, regulation, and inspection to participate in the discussion and assess the information presented.

The presentation and discussion with the FDA broke new ground in proposing an approach to industry best practices, which included implementation of an HACCP program, to ensure the safety of the canned ackee fruit. This, along with the demonstrated commitment and knowledge of the industry, Jamaican technical experts, and the regulator – the BSJ – gave the FDA the assurance it needed to be able to move the process forward. At the end of the initial discussions on the matter, the FDA agreed to revisit Import Alert 21-11 on Ackees, Canned in Brine, if Jamaica were able to provide assurances that the GOJ, through the BSJ, would:

- Require HACCP programs for all firms exporting to the United States
- Not issue an export certificate to firms not in compliance with HACCP, US Good Manufacturing Practices (GMP), and LACF regulations
- Develop a sampling plan for FDA compliance purposes and implement it.

The FDA agreed to a process that would involve further information being provided to them in the form of a detailed technical review paper, a review of the BSJ's regulation of the industry, and inspections of the factories in Jamaica selected to be among the first to be considered for exporting to the United States. In the spirit of the partnership being developed, the FDA also provided assurance that it, in turn, would:

- Establish a maximum safe level of consumption of hypoglycin based on an assessment of risk
- Identify and validate an analytical method to measure hypoglycin levels in the fruit
- Provide technical assistance in the form of training Jamaican laboratory scientists to ensure they are capable of testing ackee using the FDA's analytical method

- Conduct an informal audit of the laboratories to ensure analytical capability in testing ackee products to give the FDA confidence in the analytical packages submitted[2]
- Review and comment on HACCP programs submitted by processing facilities
- Inspect Jamaican ackee processing plants (GMP, LACF, HACCP), and provide technical assistance on how to improve GMP, HACCP, and LACF processes to be in compliance with US regulations
- Provide ongoing training to BSJ inspectors in FDA food regulations (GMP, LACF, HACCP) and how to assess compliance.

Once this process was satisfactorily implemented and the assurances on both sides delivered, the FDA undertook to exempt Jamaican firms that were in full compliance with HACCP and US regulations from the import alert. In essence then, through the instrumentality of the US ambassador to Jamaica, the US government undertook to facilitate the achievement of the major goal of opening the US market to processed ackee products from Jamaica while ensuring that the products were safe for human consumption. The FDA also sought to achieve another subsidiary, but very important, goal: developing a working relationship with the competent authority in Jamaica (the BSJ) and providing training to their inspectors in order to ensure continued oversight of ackee processors (Saltsman and DeVlieger, 2012).

6.4 IMPLEMENTATION

Prior to the mission to the FDA, the JEA was approached as to whether it would be willing to explore the possibility of recommencement of ackee exports to the United States. Dr. André Gordon of Technological Solutions Limited (TSL), an applied food science and technology services company, was asked to lead the process on behalf of the JEA to develop an implementable approach to getting Jamaican canned ackees back into the US market. Having previously convinced key colleagues at the JEA that it would be possible to get the FDA import alert lifted if it were possible to get through to the FDA for a formal hearing, Dr. Gordon

[2]It must be noted that, at the time, there were (and remain) no U.S. laboratories, outside of FDA, that can test ackee for hypoglycin. All ackee-processing firms, regardless of what country they are located in, must send their products to Jamaica for testing.

was given the lead role as chair of the initial working group. His confidence arose from having dealt successfully with both the British and the Canadians in having their scientists and regulators accept the safety of Jamaican canned ackees (Gordon, 1995a,b). Meetings were held with the US embassy and the USDA,[3] out of which a schedule of activities was determined. A proposed timeline was developed and the JEA-led Jamaica Ackee Task Force (JATF) was formed. The membership of the task force is shown in Appendix II.

As part of the overall program of regaining market access, the JATF developed and presented a schedule of activities and a budget, which was subsequently approved and funded.[4] Based on the advice of the chair, the task force accepted and mandated that HACCP had to be implemented by the ackee industry. This approach required those companies seeking to export ackees to the United States to transform from a quality-based to a food-safety-based process, including becoming HACCP compliant. A technical team of food safety experts from TSL assessed several firms and selected four for support in implementing HACCP-based food safety systems. The four firms were Canco Limited (St. Thomas), Ashman Food Processors (St. Catherine), Tijule Limited (Clarendon), and West Best Limited (Westmoreland). By March 1999, assessment and upgrading food safety standards in these firms had commenced, initially focusing on training in basic GMPs. Modernizing the food safety infrastructure in the four firms intensified significantly following a successful meeting with the FDA/CFSAN in Washington, DC. At that meeting, the JATF presented the FDA with a compelling, scientifically valid package to show that Jamaica had the data, cultural practices, and scientific underpinning to consistently produce safe canned or frozen ackee products.[5] This was the presentation delivered to the FDA's staff of horticultural specialists, toxicologists, food safety specialists, and regulators (Gordon, 1999a).

[3]Mrs. Yvette Perez, the USDA resident representative, was a key facilitator of the process.
[4]The program cost approximately US$350,000 and was financed by the National Development Bank (for the processors) and grants from the Ministry of Agriculture, the Technology Innovation Fund of the National Commission on Science and Technology, and the USDA.
[5]The process was led by Dr. Gordon, supported by his team at TSL and working closely with Mr. James Kerr, the chief chemist at the BSJ, who played a critical role in the process.

Table 6.2. Key Members of the FDA Team Implementing the Jamaica Ackee Outreach Program

Name	Title	Division
Dr. Terry Troxell	Division Director	Office of Plants, Dairy Foods, and Beverages, Center for Food Safety and Applied Nutrition (CFSAN), FDA
Dr. Joyce Saltsman	Project Lead	Office of Plants, Dairy Foods, and Beverages
Dr. John Vanderveen	Office Director	Office of Plants, Dairy Foods, and Beverages
Dr. Cathy Carnevale	Director	Office of Constituent Operations, CFSAN
Dr. Lois Beaver	Multilateral Programs Desk	International Affairs Staff, FDA
Ms. Debra DeVlieger[1]	Lead Investigator	Office of Regulatory Affairs, FDA
Mr. George Ware[1]	Chemist	Southeast Regional Laboratory, FDA

[1]Both Ms. DeVlieger and Mr. Ware became involved later on, in 1999 and 2000, respectively, when the actual mission to Jamaica for the inspections and determining the suitability of the analytical method came to the fore.

In response to the meeting with the Jamaican delegation, the FDA formed a food safety team to develop a program that would help determine whether an ackee firm could meet FDA requirements in order to be listed as exempt from the ackee import alert (IA 21-11). They also established the Jamaica Ackee Outreach Program that was managed by the food safety team (Table 6.2) and sought to provide guidance and technical assistance to firms to help them comply with the program. Through this program, which included a process that all parties agreed to, the FDA was able to establish a working relationship with the inspectorate of the Bureau to strengthen the Bureau's inspection and regulatory capabilities. In addition, based on the discussions in Washington and their own assessment (J. Saltsman, 2004, FDA Office of Plants, Dairy Foods, and Beverages, personal communication), they established a working relationship with the leading food safety expert and industry spokesperson on ackees in Jamaica.[6] This program of bilateral cooperation was instrumental in building the modern ackee-processing industry that exists today.

On return to Jamaica, the JATF immediately started working with the four firms to ensure they could meet FDA requirements, allowing 5 months to do so. TSL's team[7] started preparatory work with the four

[6]Dr. André Gordon of Technological Solutions Limited.

[7]This included Dr. André Gordon, Ms. Veronica Morgan, and Mr. George Blake.

firms, starting with basic GMP training and introduction to HACCP systems. TSL also undertook significant capacity-building at the JBS, which included extensive training of the Bureau's inspectors in conducting GMP inspections to determine if a firm was in compliance with both JBS regulations and FDA regulations (Code of Federal Regulations (CFR) Title 21, Parts 110 and 113). This was followed with more detailed training for the JBS inspectors on food safety and HACCP, to clarify the difference between HACCP-based food safety systems and the quality-systems-based (ISO-focused) Jamaican regulatory approach that was in effect at the time. Specific training for the inspectors included "HACCP Concepts and Principles" (August 1999) and "Auditing of HACCP Systems in Jamaican Factories" (September 1999). More detailed training of each of the selected factories focused on providing hands-on support with documentation of pre-existing good practices. Critically, many of the processes for racking, handling, and processing ackees (Figures 6.1 and 6.2) were improved to ensure full compliance with 21 CFR 110 and 113, and also to improve the efficiency of the operations.

The JBS made several important regulatory contributions to the process of preparing ackee factories for the US export market. First, they assumed responsibility for preparing an official process standard for canned ackee in brine, which required application of HACCP principles. TSL, working closely with the Bureau, drafted the base standard.[8] The standard (JS 276) was approved by the minister in October 1999, prior to the visit of the FDA. In addition, the Bureau confirmed that HACCP programs implemented in each factory were fully compliant with the standard and provided analytical services, along with TSL, in the area of hypoglycin testing.

[8]At the time, TSL were the only people in Jamaica comfortable with HACCP-based food safety systems, of which Dr. Gordon was the pioneer and leading proponent. HACCP implementation and acceptance as an effective tool to ensure food safety was still in its infancy in the United States at the time, with only a seafood HACCP (21 CFR 120) on the books, and some HACCP requirements being demanded in the meat industry by the USDA Food Safety and Inspection Service. Canada had started implementation through the Food Safety Enhancement Program and the Europian Union had regulations based on HACCP for meat, seafood, and dairy products.

(a)

(b)

Fig. 6.1. Rack-ripening of ackees: (a) traditional approach – open-air rack-ripening on wooden racks; (b) postintervention – ripening on covered metal racks.

(a)

(b)

(c)

Fig. 6.2. Postintervention processing: (a) husk removal; (b) cleaning; (c) filling of cans. (Source: A. Gordon, 1999.)

At the time, formal filing of thermal processes specific to each factory, as well as the validation of processes at each plant and verification of vent schedules and temperature distributions for each retort, was not always being done routinely in Jamaica. Subsequent to the visit to Washington, based on the queries raised, extensive work had to be done on thermal processes, including the development and/or validation of vent schedules and thermal processes for all of the four plants selected for support.[9] These scheduled processes would then have to be filed with the FDA once their inspections were completed. Based on a final assessment by TSL before the FDA came to Jamaica, it was decided that only three of the four factories had enough of the systems in place to meet the FDA's requirements, and so only Tijule Limited, Ashman Food Processors Limited (AFPL), and Canco Limited went forward for inspection.

Between October 26 and November 5, 1999, facilitated by TSL, the FDA representatives came to Jamaica to inspect the three establishments (Figure 6.3). Their inspection included detailed assessment of each factory's LACF process to determine compliance with 21 CFR 113, including the compliance of retorts and thermal process systems (Figure 6.4); an evaluation of the firm's implementation of GMPs; and a detailed assessment of the firm's protocols for controlling hypoglycin. Of the three firms, Canco Limited and AFPL demonstrated good control of their thermal processing operations and the safety of the ackees by effective control of hypoglycin. The FDA also visited the Bureau's Chemistry Laboratory and held discussions with the analysts about the analytical method for hypoglycin.

Because of concerns for public health and the toxin potentially present in the fruit, the FDA's Office of Plants, Dairy Foods, and Beverages in CFSAN raised the question about a safe level of HGA consumption. Jamaica had already established a safe limit for HGA in canned ackee of 150 ppm. This was based on 50% of a limit of 300 ppm proposed as a safe level by Bates (1991) and data from rat feeding studies (BIBRA, 1995). Jamaica also presented these data and other information

[9]Thermal processing work was led by Dr. Gordon, a thermal process authority, using standard E-Lab thermocouple systems (from E-Lab) as well as the then new Data Trace cordless probes (from Mesa Labs, CA) and CTemp software (from Campden and Chorleywood Food and Drink Research Association).

Fig. 6.3. The FDA inspectors with the USDA resident representative and TSL support team. Left to right: Jeffrey Brown (consumer safety officer, FDA CFSAN), Dr. Joyce Saltsman, team leader (interdisciplinary scientist, FDA CFSAN), Yvette Perez (USDA resident representative), Veronica Morgan (director, TSL), Dr. André Gordon (managing director/principal consultant, TSL), Debra DeVlieger (lead investigator, FDA Office of Regulatory Affairs). (Source: A. Gordon, 1999.)

to the Canadian and UK food safety authorities, which resulted in both countries accepting the 150 ppm HGA as a safe level (Gordon, 1995a,b). The FDA, however, conducted their own risk assessment and determined that 100 ppm would be considered the maximum safe level that would be permitted in canned or frozen ackees. This would become the level that the JBS would use to monitor compliance of the exporting firms with US requirements.

With a focus now on the analytical method for determining HGA concentration in ackees, the FDA needed to evaluate the analytical

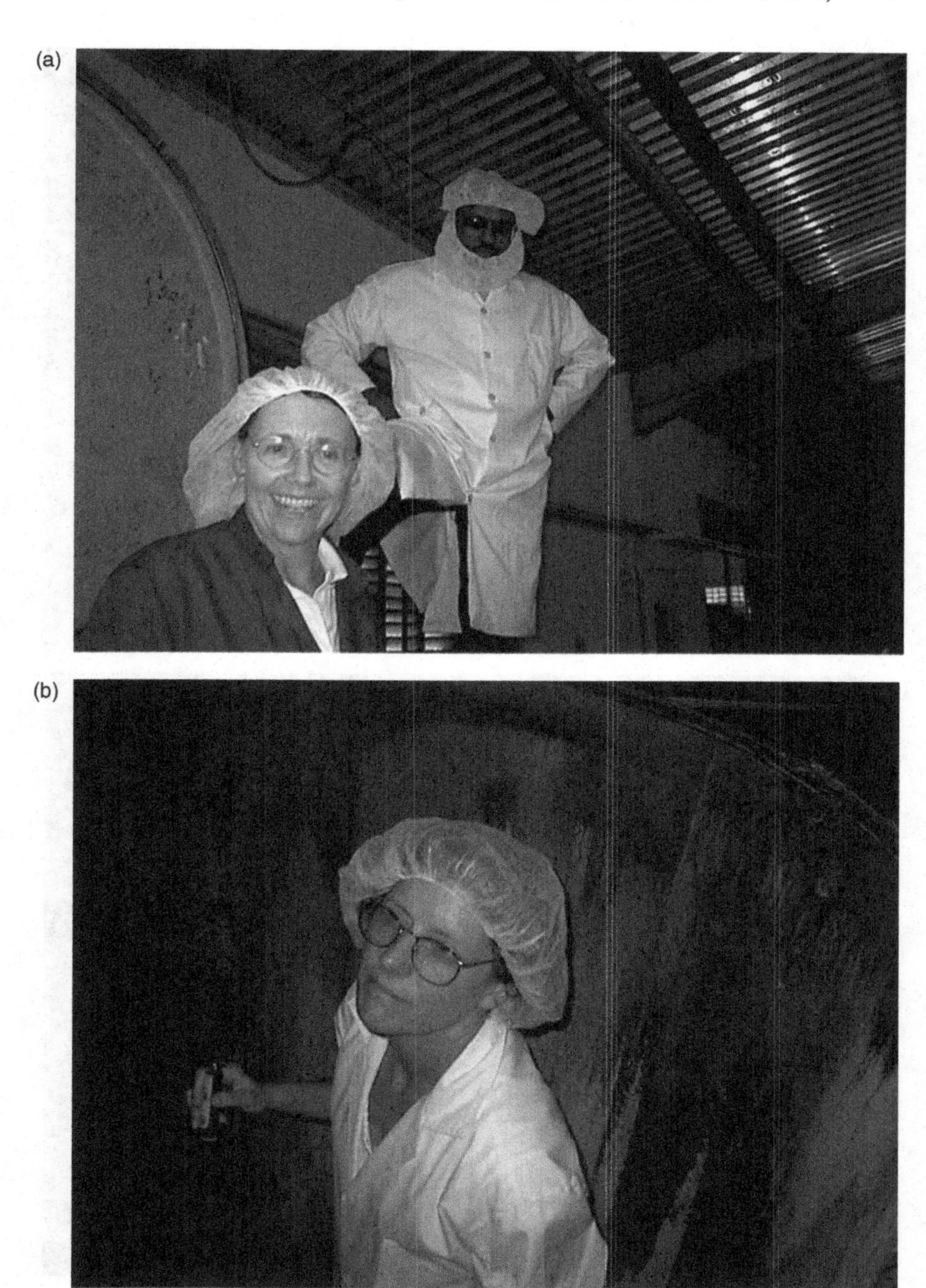

Fig. 6.4. FDA inspection of thermal process systems at Jamaican plants, including: (a, b) the dimensions of the retort, (c) the cross-sectional area of the spreader, and (d) the diameter of holes in the retort baskets. (Source: A. Gordon, 2002.)

(c)

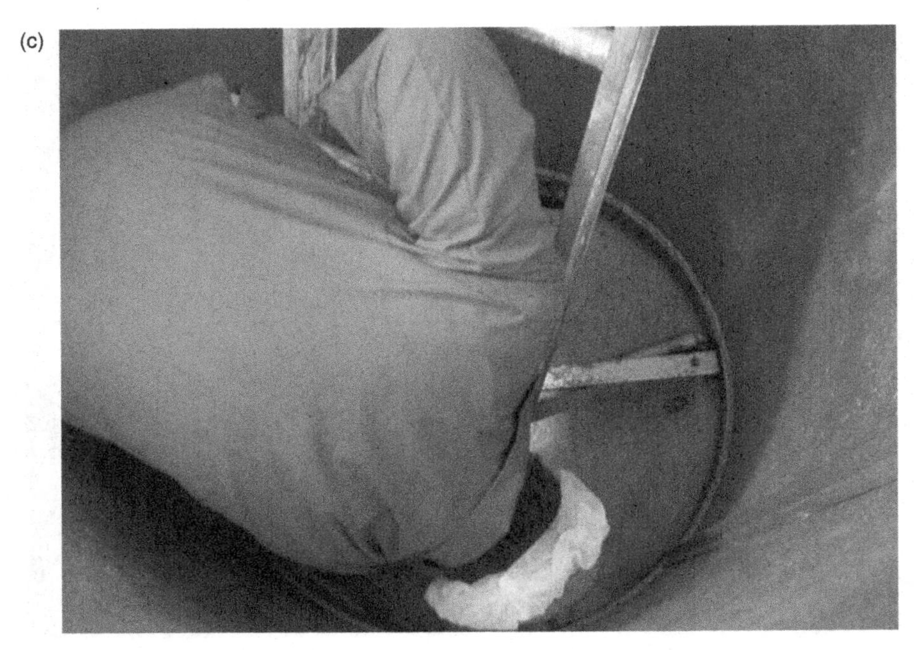

(d)

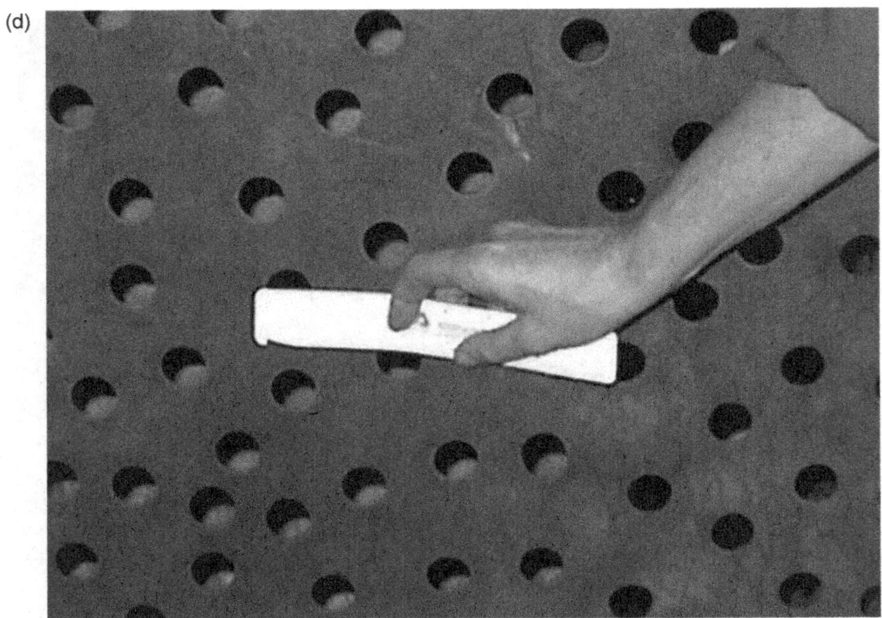

Fig. 6.4. (Continued)

accuracy of the laboratories in Jamaica that were equipped to test for the toxin. This was necessary for two reasons: (1) the FDA ackee program required ackee firms requesting an exemption from the ackee import alert to submit the results of HGA tests of several lots of product, and (2) the FDA import program for products with an import alert usually required a foreign firm's products to be tested in the United States. The problem for the Jamaican firms was that there was no laboratory in the United States that performed HGA determinations, so the FDA ackee program was designed to permit testing of the product in Jamaica. To evaluate the analytical accuracy of Jamaican laboratories, the FDA conducted a collaborative study with two laboratories (TSL's laboratory – led by Dr. Gordon, and JBS's laboratory – led by Mr. James Kerr) using the same method (Sarwar and Botting, 1994) for determining HGA (Ware, 2002). The results of the collaborative study proved to the FDA that the two laboratories were capable of performing the test accurately; therefore, all ackee firms were mandated to have five lots of product tested for toxin levels in order to gain access to the US market. On July 6, 2000, AFPL and Canco Limited were permitted to start shipping product to the United States.

Canco Limited and AFPL now had preliminary access to the US market but still needed to demonstrate that they were shipping product that did not exceed the maximum allowable level for hypoglycin (i.e., 100 ppm). This final hurdle required each firm to ship five "clean" shipments to the United States (i.e., all lots of product had to comply with the hypoglycin requirement) before being taken off the "detention without physical examination" (DWPE) listing. DWPE is the process by which all incoming shipments of potentially violative products (in this case, canned ackees) were held at the US port of entry (US FDA, 2000). Once a shipment was detained, the FDA had the authority to audit the products (i.e., test lots for hypoglycin levels) to ensure compliance with requirements. If all five shipments from a firm were in compliance, the firm was listed as an exception to the ackee import alert, so subsequent shipments could flow freely into the country. The first shipment from Jamaica to the United States was sent by Canco Limited in August 2000 (Gordon, 2011; The Daily Gleaner, 2000). Both firms were eventually listed as exceptions to the import alert, thus signaling the beginning of legal ackee trade with the United States.

6.5 ACCOMPLISHMENTS

The approach to reopening the US market for Jamaican canned ackees involved two complementary programs: the program of the JATF in Jamaica and the FDA's Jamaica Ackee Outreach Project (based in the United States). Collectively, these accomplished the goal of getting the fruit back into the United States while ensuring it was safe and met all US regulations, an achievement that has transformed the lives and fortunes of many rural folk who depend on an industry that has grown fourfold since July 2000. The process resulted in the development of a mutually respectful working relationship between FDA's Office of Regulatory Affairs (ORA), CFSAN, the BSJ, two trade associations (the JEA and the Jamaica Agro Processors Association (JAPA)), and industry. This remains the "gold standard" in the Caribbean for collaboration between industry, government, and bilateral partners, and can be a template for future such efforts.

Other important accomplishments of the process include:

1. A revitalization and transformation of the agroprocessing sector in Jamaica. Ackees, as the base of small agroprocessing in the country, by 2015 not only earn over US$15 million per annum (up from US$4.4 million in 1999) but have shown that all levels of the food industry can implement and profitably operate GMP and HACCP-based systems to consistently produce safe, quality foods. This is therefore becoming the norm in the food industry in Jamaica and is spreading throughout the Caribbean region.
2. A catalyzing of the establishment of HACCP-based food safety programs. This served as the basis for the production and export of food from Jamaica, which was the first emerging market to establish such programs.
3. The successful training of BSJ's inspectors in HACCP and LACF by the FDA's national food expert, such that they could assist during FDA inspections and, on their own, inspect ackee firms for compliance with US HACCP, GMP, and LACF requirements. The BSJ has now expanded its inspection responsibilities to do more HACCP and LACF inspections in other types of processing plants in Jamaica.
4. Transformation of the regulatory underpinning of the food industry in Jamaica (and the Caribbean Community – CARICOM) region.

Jamaican Standard (JS) 276 and several subsequent voluntary and mandatory standards and regulations now exist that require GMP and risk-analysis-based food safety controls and impact the entire food chain and their export markets.

5. The setting of food safety standards in Jamaica by ackee processors for other processing firms. In the early phase of the project, the FDA inspected four processing firms and found violations in LACF requirements. The FDA Team Lead (Dr. Saltsman, CFSAN's Office of Plants, Dairy Foods, and Beverages (now the Office of Food Safety)) and National Food Expert (Debra DeVlieger, ORA's Division of Field Investigations), along with TSL's team, worked with the firms to help them understand the HACCP and LACF requirements, their existing weaknesses, and what would be needed to correct them.[10] With the Food Safety Modernization Act (FSMA) on the horizon, the ackee processors are being used as examples to all food processors in Jamaica and throughout the Caribbean region to show that "it can be done."

In addition to the important structural, international relationship, and industry-wide accomplishments of the process, signal achievements were also made in the area of scientific understanding and management of the fruit. These were:

1. The identification and refinement of an analytical method for hypoglycin (HGA) analysis. Mr. George Ware of the FDA's Southeastern Regional Laboratory published the analytical method in the Association of Official Analytical Chemists (AOAC) Journal.
2. Development of a robust sampling method for assessment of canned ackees for the risk of exceeding the hypoglycin safe limit. Southeastern Regional Laboratory (SRL) and CFSAN published the results of this collaborative study.
3. Training of Jamaican laboratories in the analytical method and the FDA's reporting requirements for analytical packages. Two

[10]Initial inspections involved CFSAN's Office of Plants, Dairy Foods, and Beverages and ORA's Division of Food Investigations (DFI), but later inspection trips included OFS's food technologist (FDA team lead) and a DFI inspector (national food expert), or just DFI inspectors. Additional inspectors were later trained to perform ackee inspections and determine compliance with HACCP, GMP, and LACF regulations, and are now a mainstay of FDA Food Safety Modernization Act inspections in the Caribbean region.

laboratories were informally audited and found capable of testing ackees using the FDA's methodology.[11]

4. Completion of the FDA's qualitative risk assessment for HGA in processed ackee products and establishment of a compliance level.

6.6 COST/BENEFIT ANALYSIS

The overall program that led to the successful re-opening of the US market was not without significant costs to the United States and Jamaica. The FDA spent approximately US$20,000 per visit, which included visits to Jamaican firms and a visit to an ackee firm in Haiti (Saltsman and DeVlieger, 2012). The first two visits focused on the inspection of firms for approval for exception from the ackee import alert, and training, in collaboration with TSL, of BSJ inspectors in HACCP and LACF. Additional trips focused on inspecting other firms that are now listed as exceptions to the import alert (a total of 12 firms, representing three countries) and additional training for BSJ inspectors.

In Jamaica, the extensive program described previously cost approximately US$350,000 and was funded by a combination of stakeholders in the public and private sectors, described previously. This covered the research program on HGA and ackees, training in HACCP for the processors and regulators, and systems implementation for HACCP-based food safety systems for the processors. This excludes investments in individual plants, which ranged from approximately US$80,000 to US$420,000 at the time (A. Gordon, TSL, unpublished data).

The main outcome of the combined collaborative effort between the FDA and the BSJ was the successful re-entry of canned ackees into the US market. From the FDA's perspective, the program accomplished its goal of opening the US market to processed ackee products from Jamaica, while ensuring that the products were safe for human consumption. Other significant outcomes, listed previously, include the building of a solid relationship between the FDA, the BSJ, and the industry; the provision of assurances of the BSJ inspectors' capabilities to inspect processors according to FDA requirements and regulations; the scientific

[11]Initially (in 2000), only the BSJ and TSL were approved for testing. Subsequently, the Mona Institute of Applied Sciences was also approved.

information derived; and the major and sustainable transformation of the ackee processing sector. The significant benefits from the ackee program continue to be felt by the US government, even as it rolls out the FSMA activities and requirements for exporters throughout the Caribbean and elsewhere. Jamaica, on the other hand, has received handsome dividends of the transformation of the trajectory of its rurally based agroprocessing sector and a greater than fourfold increase in export earnings over a 12-year period. By any measure, the efforts of the Jamaica Ackee Outreach Program and the Jamaica Ackee Task Force have been an unqualified success and provided exceptional returns on the funds and effort invested.

6.7 SUMMARY

Ackee, like all foods, must be safe for human consumption. This is the concern of all governments internationally who participate in world trade and was the concern of the FDA when they instituted the import alert on canned ackee in 1973. Without the science-based knowledge of the fruit, along with proper handling and processing practices and a method to effectively demonstrate a safe level of toxin in the processed product, it was not possible for the FDA to allow the resumption of trade. This meant, therefore, that for over 20 years, Jamaica's efforts to get the import alert amended to allow export of this economically important fruit were unsuccessful. Finally, in 1999 a collaborative effort between the Government of Jamaica, Jamaican industry, Jamaican scientists, and the FDA was established to address the safety issues. This involved a national effort in Jamaica, led by the Jamaica Ackee Task Force and a companion effort in the United States through the Jamaica Ackee Outreach Program.

The Jamaica Ackee Task Force had a transformational impact on the ackee industry by reinforcing good agricultural practices (GAPs) and GMPs, implementing HACCP-based preventive controls, and initiating local regulatory reform in Jamaica. This, combined with the FDA's outreach program with the BSJ and ackee industry, met the goal of enabling Jamaica to again export ackee to the United States (beginning in 2000). The program had the added benefit of improving the GOJ's understanding of US regulatory requirements and its ability to assess the food industry for compliance with those requirements. The approach taken was, therefore, not only successful, but has had several lasting benefits for both Jamaica and the United States. Moreover, it presents a template for how market-access challenges for traditional fruits may be successfully addressed.

CHAPTER 7

Dealing with Trade Challenges: Science-Based Solutions to Market-Access Interruption

André Gordon[1]
[1]Technological Solutions Limited, Kingston, Jamaica

ABSTRACT

A deterioration in practices, controls, and other factors caused hypoglycin A (HGA) levels in Jamaican ackees to exceed the regulatory limit and resulted in the reimposition of an import alert on the product in December 2005. The Jamaican industry, the government, and the US Food and Drug Administration (FDA) collectively worked to correct the problem and prevent a recurrence. This included collaborative research on the cause of the problem, developing and implementing sustainable solutions, retraining the industry and the Bureau of Standards Jamaica (BSJ) inspectors, and developing an isolation and testing mechanism through which compliant product could meet the FDA's requirement of five consecutive nonviolative shipments. The congruence and accuracy of the analytical regime being used for ackees by the FDA, BSJ, and Technological Solutions Limited was also reconfirmed and a statistically

Food Safety and Quality Systems in Developing Countries. http://dx.doi.org/10.1016/B978-0-12-801227-7.00007-X

sound sampling regime was developed. The collaborative approach to the ackee export trade between Jamaica and the United States contributed to industry-wide implementation of food safety systems and improved competitiveness, which continues to benefit Jamaica and can be a template for other trading partners facing similar challenges.

Keywords: breakdown in practices; ackee; hypoglycin A; import alert; market access; analytical regime; sampling; HACCP

7.1 INTRODUCTION

During the course of ongoing trade between countries, challenges often arise that can lead to market-access interruptions, the length and severity of which will vary depending on the underlying cause for the discontinuation of trade. Sometimes these challenges arise from tariff- and/or quota-related issues, but most often the cause is an unexpected or unforeseen circumstance that causes the importing country to have doubts about the safety of the exporting country's goods or its compliance with the importing country's regulations. In these cases, the issue of technical barriers to trade (TBT) may arise, as the concerns leading to the market-access interruption are typically technical or science-based in nature. These may include noncompliance with labeling requirements; the use of nonapproved ingredients, processing, or production aids (including pesticides, coloring agents, and other ingredients); or, as has been the case with Jamaica's ackee exports to developed-country markets, uncertainty about the scientific basis on which the food could be accepted as safe. Where noncompliances or problems arise from technical or scientific issues (as for the cases in Figure 7.1), it is important to have the knowledge or expertise to be able to effectively mitigate these problems to the satisfaction of the importing country's competent authority (CA).

In the case of TBT that are related to labeling, ingredients, production input, or practice, a solution to the problem can usually be found in a relatively short time if the exporter and the CA, or the respective CAs in the exporting and importing countries, agree on how the matter should be handled. The exporter, alone, through the use of technical support organizations and/or with the support of their CA, then demonstrates with the use of appropriate scientific or technical documentation

(a)

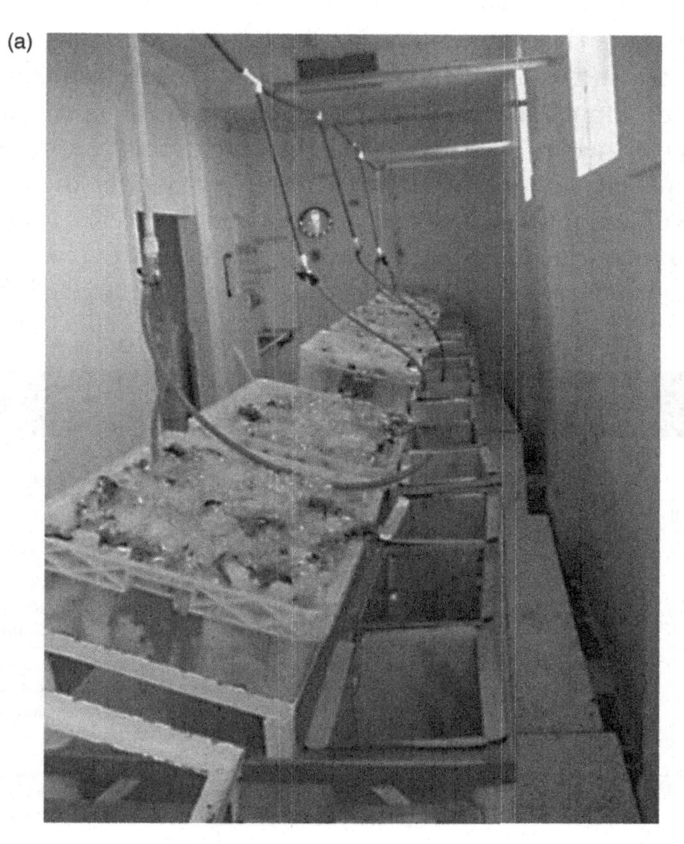

(b)

*Fig. 7.1. Processing of two products that faced market-access challenges: (a) queen conch (*Strombus giga*) and (b) canned processed cheddar cheese.* (Source: A. Gordon, 2001.)

that the matter has been addressed to the satisfaction of the importing country. In the case of a labeling breach, the label is amended. For production inputs, ingredients, or additives, the producer can demonstrate, supported by their local CA, if required, that the source of the noncompliance has been removed from the product or the production practice discontinued. Some technical noncompliances that have arisen for exports from the Caribbean and how these were addressed are shown in Table 7.1, and can serve as templates that can be applied in similar

Table 7.1. Curtailment of Caribbean Exports to Developed-Country Markets Due to Technical and Scientific Issues, and Their Mitigation

Product	Exporting/Importing Country	Nature of the Issue/ Challenge	Approach to Resolution
Canned callaloo (canned *Amaranthus dubius*)	Jamaica/the United States	Unapproved pesticide residues found	Worked with the USDA to change the pesticide to a US-registered one; worked with farmers to implement proper postharvest interval management, implement traceability, and change field husbandry practices
Crackers	Trinidad and Tobago/the United States	Noncompliant labeling/ proof of certification of FD&C colors	Amended label to become compliant; obtained FDA certification of all batches of FD&C colors used in the food; produced as evidence – kept on file
Red pepper sauce	Jamaica/the United Kingdom	Disallowed color (Sudan red) found in product	Source of banned dye identified by analyses as being admixed Central American pepper mash; this was removed
Queen conch (*Strombus giga*)	Jamaica/the European Union	Unapproved CA; nonequivalent regulatory framework; exporters not GMP and HACCP compliant	Upgrade of the CA, development and passage of equivalent legislation; implementation and monitoring of GMP and HACCP at conch processors
Mango kuchela (*Mangifera indica* sauce)	Trinidad and Tobago/the United States	Product not filed with CFSAN[1]; formula safety of product not proven	Shelf stability and safety of product established through studies and thermal process validated and filed with CFSAN's LACF division
Sea (Irish) moss (*Gracilaria* drink)	Dominica/the United States	Thermal process not defined; product not filed with CFSAN	Thermal process developed and validated and filed with CFSAN's LACF division
Canned processed cheddar cheese	Jamaica/the United States	Product not filed with CFSAN; formula safety of product not proven	Product established as a formula safe dairy product by studies (using the *Tanaka Principle*) and thermal process validated and filed with CFSAN's LACF division

[1]The United States Food and Drug Administration's Center for Food Safety and Applied Nutrition (CFSAN).
Source: Technological Solutions Limited (TSL), the author's company, which has handled all of these matters.

circumstances. More details on some of these and other food-safety- and quality-related issues in developing countries can be found in Volume II of this series.[1]

Where the problem has to do with inadequate scientific evidence of the safety of the food item being exported or proof of its compliance with technical requirements (e.g., validation of the effectiveness of a production process or the equivalence of a different, traditional approach to assuring safety), the matter becomes more complex and can result in a protracted cessation of trade. This was the case with the import alert on canned ackees from Jamaica going into the United States from 1973 to 2000, as was examined in Chapter 6. As more countries seek to supply, first, their diasporas in developed countries, and then the general population with popular traditional foods – in keeping with a growing trend toward more healthy traditional foods and experimentation with exotic foods (International Food Information Council (IFIC) Foundation, 2013; Sloan, 2014) – this situation will arise much more often. This chapter therefore seeks to use the circumstances surrounding other market-access challenges that have arisen with the export of ackees to developed-country markets and the solutions applied, to highlight and demonstrate some possible options for exporting entities faced with similar technical and scientific challenges to continued trade. It presents issues that arose with exports to Canada, the United Kingdom, and again with the United States, and examines how these were addressed.

7.2 CANNED ACKEES EXPORTED TO CANADA: MITIGATING MARKET-ACCESS PROHIBITION

Canned ackee has been exported to Canada from Jamaica since the 1980s and, more recently, Canada is also importing the product from Côte d'Ivoire and Haiti. In late 1994, however, the Canadian authorities became aware of, and increased their interest in, the ackee being imported from Jamaica (the only exporter of the product to Canada at the time) due to a reported illness associated with ackee consumption in Canada (Gordon, 1999a,b; F. Jamieson, Health Protection Branch, Health and Welfare Canada, 1995, personal communication). Health

[1]Food Safety and Quality Systems in Developing Countries, Volume II: Case Studies of Implementation.

and Welfare Canada (now Health Canada), the body responsible for the safety of food in Canada at the time,[2] had concluded, based on the information available to them, that the levels of HGA in the product associated with the reported incident were high enough to cause illness. No other information on the case was provided, making it difficult to further evaluate the report. However, when assessed by the Jamaican team dealing with the crisis at the time[3] against the scientific information available on the fruit, its toxin, and the levels associated with Jamaican vomiting sickness (JVS), the data did not support the validity of the claim. The levels of HGA found in the cans were well within the range known by the Jamaica Bureau of Standards (JBS)[4] to be routinely present, and that previously reported for canned ackees (Brown, 1989; Chase et al., 1989), none of which had ever been associated with illnesses.

The Canadian authorities requested and were furnished by the JBS with information on ackees, including data on HGA levels obtained from periodic testing of canned ackees for export, and the preventative systems in place to ensure that only safe, compliant canned ackees were produced and shipped, including the program for monitoring the producers. Also requested were information on the analytical protocol being used and HGA standard material to be used by the Canadian authorities in their own monitoring program. It was implied that failure to comply with these requests might result in Jamaica's loss of this very important market for the product. The assessment of the risk based on the reported HGA content of the cans, along with other technical and scientific data on ackees and HGA, was detailed in a review paper (Gordon, 1995a) that accompanied Jamaica's submission to Canada in response to the reported incident. The interaction with the technical professionals in Jamaica and the data and information provided, along with a sample of the HGA standard being used by the JBS, were sufficient to satisfy Health Canada. This incident showed the importance of undertaking scientifically valid studies to gather data on traditional foods being exported to third-country markets, as well as analyzing and collating available information and having

[2]This made them the competent authority (CA) from the perspective of trade.
[3]This consisted of Dr. André Gordon, then of Grace Technology Centre, Grace Kennedy & Co. Ltd., and the Jamaica Bureau of Standards, through Mr. James Kerr.
[4]This name has been changed to the Bureau of Standards Jamaica.

the technical competence to be able to demonstrate the safety of such foods to importing markets.

7.3 PREVENTING MARKET-ACCESS INTERRUPTION FOR CANNED ACKEE EXPORTS TO THE UNITED KINGDOM

In 1995 the United Kingdom also began closer scrutiny of ackees after the issue of the potential hazard from hypoglycin had been brought to their attention by a visiting FDA official, who had become aware that the United Kingdom was importing an item that the United States had banned for safety reasons. This was exacerbated when a major retail chain, J. Sainsbury p.l.c., removed the product from its shelves, also for possible safety concerns, having become aware that improperly handled and processed product might pose a risk to consumers. Sainsbury's actions were likely driven as much by the company's reputation for putting the consumer first as they were by recent changes in food laws in the United Kingdom, which transferred the responsibility for some aspects of food safety from the regulators to the retailers. The United Kingdom was at the time the major market for Jamaican canned ackees, also being the market that gave the best returns to exporters and, as such, was critical to the viability and sustainability of the ackee processors and exporters and the sector as a whole. Consequently, when the then Ministry of Agriculture, Fisheries and Food (MAFF) food safety function, now the Food Safety Agency (FSA), indicated to Jamaica that their concerns about the safety of canned ackees being imported into the United Kingdom were such that they were contemplating placing a ban on all ackee imports, Jamaica and its exporters had to react quickly.

MAFF had indicated through the Jamaican embassy in London that they were seriously considering restricting importation of canned ackees pending a review of the product's safety status. Discussions ensued between the UK government, through MAFF, and the Jamaican embassy and other Jamaican government officials to mitigate the situation. During the process of those discussions and others taking place between the BSJ, major importers, and major exporters from Jamaica, it became apparent that MAFF was being driven to act because, among other factors, there was a dearth of scientific and technical information on the fruit and its safety available to them. This was pre-empted when a detailed technical presentation was made to a meeting of a multidepartmental scientific and technical team at MAFF, providing information on the product, the nature of the illness, the toxin, and the systems in place to ensure safety.[5] This was followed by a review paper on the issue (Gordon, 1995b). MAFF were satisfied that there was no undue cause for alarm, the sale of the product continued uninterrupted, and the major retail chain relisted the product shortly thereafter.

7.4 REIMPOSITION OF THE US IMPORT ALERT: LIFTING THE BAN, AGAIN

The inability of Jamaican exporters to supply the US market with canned ackees and the resulting reaction of the market of increasing the lucrativeness of the trade through increased prices led to many new entrants coming into the industry by 2005. This led to a fragmentation of the industry and reduced cooperation among processors and exporters, with some beginning to deviate from the practices that had been established to facilitate the lifting of the import alert in the United States (Martin-Wilkins, 2005). The BSJ, faced with greater difficulty in regulating the industry, had challenges in ensuring that only those ackees produced in keeping with scientifically supported, hazard analysis critical control points (HACCP)-controlled production made it into the US market. However, this was not immediately conveyed to the FDA as the BSJ sought to address the matter locally in Jamaica. Among the more worrying practices was the move by some processors toward buying and using only opened ackees, which saw them losing control to some extent over the ripening and selection processes, which are critical to ensuring the safety of the fruit (Gordon, 2011). As such, the handling and selection processes being used by some participants in the industry were no longer documented or validated by scientific studies of the hypoglycin content of the fruit at different stages. These changes in the industry may have coincided with changes in the way fruits ripened in response to the severe stress that trees had undergone during the passage of a major hurricane in 2004. Whatever the actual cause, the resultant HGA profile of the fruits processed in most plants across the country changed, resulting, in some cases, in significantly higher HGA levels (Bliss, 2008).

[5]This was done by Dr. André Gordon, then of Grace Technology Centre, Grace Kennedy & Co. Ltd., on behalf of Jamaica and the Jamaica Exporters' Association.

In the United States during 2005, as a result of intelligence the FDA's Office of Food Safety (OFS)[6] received about potential unacceptable processing practices and elevated toxin levels in finished product, a monitoring program was initiated to test all incoming ackee products for toxin levels (Saltsman and DeVlieger, 2012). The monitoring program showed that ackees from all of the plants exporting to the United States were, at times, exceeding the established limit (100 ppm HGA). This resulted in all companies being put on detention without physical examination (DWPE).[7] This precipitated a crisis in the industry in Jamaica that saw the Jamaica Exporters' Association (JEA) and the Jamaica Agro-Processors Association (JAPA) approaching the FDA and United States Department of Agriculture (USDA) for assistance in addressing the challenges and getting the product back into the United States, in a process to be led by Technological Solutions Limited (TSL) (Martin-Wilkins, 2005). This involved TSL undertaking research in collaboration with the FDA and the BSJ to identify the cause of the systemic increase in HGA levels across the industry. It also required a reverification of the reliability of the analytical results being generated by all laboratories doing analyses for HGA, and a recalibration of the commercial production practices and processes, as well as the regulatory oversight of the industry.

As part of the mitigation process, the FDA undertook further training of the Bureau's inspectors in conducting HACCP and low-acid canned food (LACF) inspections. The trainers were the FDA's national food expert – Ms. Debra DeVlieger, from FDA Office of Regulatory Affairs (ORA), and the food safety expert from TSL – Dr. André Gordon. Other government representatives attended, in addition to exporters, JAPA members, and ackee processors. The program also gave the BSJ's inspectors the opportunity to participate actively in the plant inspections and gain valuable experience on top of what they learned during training. On another front, the FDA's OFS worked with the BSJ and TSL to successfully identify the cause of the violative ackee products in 2005, based on the research undertaken. A major reason for the problem was the changes in harvesting, ripening, and selection practices that had resulted

[6]Formerly the Office of Plants, Dairy Foods, and Beverages.

[7]A process by which targeted food items are automatically detained by the FDA until determined to be acceptable for importation.

Fig. 7.2. A naturally mature, naturally opened ackee (center) with two opened but immature ackees on either side. (Source: A. Gordon, 2006.)

in opened but immature fruit (Figure 7.2) becoming commonplace in the raw material supply to most processors (Campbell, 2006), including even the most vigilant processors, largely because they were unaware of the effect of the practice. Further, even where processors were aware of the practice, receivers and selectors were not able to consistently distinguish between fit and unfit (immature) open fruit (Gordon and Lindsay, 2007). The outcome of the tripartite collaborative research was suggested changes to each processor's food safety program, including stricter controls and training for ackee suppliers and changes in how fruits were selected for processing. Those companies that were successful in regaining access to the US market made the required changes and their product came back into compliance with the HGA limits.

Having settled the cause of the elevated HGA levels, a way now had to be found for the FDA to be able to assure itself of the compliance of all exporters offering ackees for import to the United States. To address this, the OFS, with concurrence from ORA and the FDA Department of

International Programs and with the cooperation of the USDA attaché in Jamaica, set up a one-time program that permitted the Jamaican companies to send their shipments to the main Jamaican international airport, where they were locked down by the USDA. At this point, the processor no longer had control of the shipments and the USDA held them until they had been sampled and tested in Jamaica. The analytical packages were then reviewed by the FDA, who gave permission to release the shipments for air freight to the United States if they were found to be compliant. The OFS also contacted the USDA agricultural statistician (Dr. Thomas Whittaker) at NC State University Agriculture Research Center for assistance in developing a sampling plan for ackee. The sampling plan greatly improved the FDA's ackee program, made it consistent with other sampling plans for natural products (Vargas et al., 2006), and overcame significant sampling bias associated with the method the FDA had used at the start of the ackee program (Whitaker et al., 2007). All Jamaican laboratories now use this sampling plan. The final aspect of the program was to ensure that all laboratories undertaking HGA analyses were getting comparable results and that the results were accurate. This was confirmed by a series of collaborative studies carried out between the FDA's Southeast Regional Laboratory (SRL), the BSJ, and TSL (Gordon, A., Lindsay, C., Kerr, J. & Ware, G. (2006)). Collaborative inter-laboratory analysis of hypoglycin A in canned Jamaican ackees (*Blighia sapida*) at different stages of maturity. Unpublished results).

Overall, the program to address the market-access interruption was very successful and appreciated by the Jamaican authorities, the JEA, and the industry, with whom the FDA and USDA worked closely to have the process designed and implemented. The detention of all ackee exports from Jamaica, which spanned over a year, cost the country over US$5 million in export earnings, a loss that created an economic burden for a significant part of the agroprocessing sector and the country. The Jamaican government and industry, in an effort to prevent a recurrence, sought to strengthen their relationship with the US authorities and established a collaborative program to manage the industry (key participants in the program are shown in Table 7.2). This program saw the FDA building the capacity of the BSJ over the ensuing years, as well as strengthening its relationship with the industry and industry experts in Jamaica and providing technical assistance, where relevant.

Table 7.2. Key Players in the Collaboration to Reopen the US Market to Jamaican Ackees in 2006

Organization[1]	Role
Jamaica Exporters' Association	Led the process on the Jamaican side; lobbied; funded and coordinated efforts with the FDA, TSL, JAPA, BSJ, and USDA
Jamaica Agro-Processors Association	Brought the issue to national attention; mobilized serious processors to comply; provided support to the JEA with lobbying; collaborated with the JBS, FDA, and TSL; funded TSL's research
Technological Solutions Limited	Provided technical leadership and coordination of overall process; undertook critical research to identify the cause of the problem; proposed solutions; provided laboratory support (HGA testing) and training in LACF for industry and JBS inspectors (with the FDA); systems implementation (for industry)
Jamaica Bureau of Standards	Had regulatory oversight (made inspections); collaborated with TSL and the FDA; provided laboratory support (HGA testing)
FDA Office of Plants, Dairy Foods, and Beverages	Led FDA's program of support; coordinated the development of solutions to the HGA and Jamaican industry noncompliance problem; developed and led collaborative support program for Jamaica
FDA Office of Regulatory Affairs	Led inspection and training
FDA Southeast Regional Laboratory	Led analytical work; developed and recalibrated analytical program for HGA; collaborated to validate sampling methodology for ackees
US Department of Agriculture	Developed valid sampling plan for ackees; helped facilitate FDA visit; facilitated lockdown of export samples that allowed recommencement of exports
Ministry of Agriculture, Jamaica	Provided the main central government support to the JEA, the industry, and other collaborators; facilitated isolation and lockdown of export shipments at airport

[1]These organizations played the main roles in working to remove the alert on ackees in 2006.

7.5 SYNOPSIS OF THE APPROACH TAKEN AFTER THE MARKET-ACCESS INTERRUPTION IN 2005/2006

A rise in the general HGA levels in Jamaican ackees, as well as a deterioration in practices and controls, caused the levels of the toxin being detected in US samples to consistently exceed the allowable limit, resulting in the reimposition of the import alert in December 2005. The FDA and Jamaican interests separately and collectively undertook a series of targeted actions to correct the problem and prevent a recurrence. These included:

- The Jamaican industry, exporters, and regulators undertaking collaborative research on the cause of the problem to come up with an implementable and sustainable solution.
- The FDA providing technical assistance to, and collaborating with, the BSJ and TSL in resolving the problems that led to the

detention of product from all Jamaican companies on the green list[8] in 2005.

- The FDA providing technical assistance by delivering training workshops to the Bureau's inspectors on conducting HACCP and LACF inspections. Inspectors accompanied the FDA team on each of the inspections and received additional training on the job.
- The JEA, JAPA, the BSJ, and the Ministry of Agriculture collaborating with the USDA attaché in Jamaica and the FDA's Center for Food Safety and Applied Nutrition (CFSAN) and Division of Import Operations and Policy (DIOP) in developing and implementing a one-time program to assist the Jamaican companies who were put on DWPE in 2005. This involved getting their shipments isolated and tested in Jamaica to meet the criteria for five consecutive compliant shipments.[9]
- Developing and implementing a long-term ackee food safety program as the number of exporting companies in Jamaica increases. This has included a protocol that has allowed the BSJ to act on behalf of, and in collaboration with, the FDA in approving new plants for export.
- Developing a program, including access to a testing laboratory,[10] that would allow processors in other countries beginning to export to meet the compliance criteria.

7.6 ACCOMPLISHMENTS AND SUBSEQUENT ACTIONS TO MAINTAIN THE GAINS

In July 2008, Mr. George Ware of the FDA, SRL, visited Jamaica to reassess and benchmark the procedures being used for determining the HGA content of samples by the BSJ and TSL against those being used by the FDA's laboratory. At the end of the exercise, all laboratories were confirmed as being equivalent, with excellent correlation between the expected and actual results (Ware, 2008). In 2009, with the increasing number of ackee processors in Jamaica and other countries contacting the FDA for guidance on getting into the US market, OFS developed

---

[8]Firms on the green list have requested and been granted an exemption that allows them to ship a product into the USA without detention that is normally on automatic DWPE. They achieve entry onto the green list by shipping five consecutive compliant ("clean") shipments.
[9]Dr. Gordon was again central to the process and led the negotiations, collaboration, and logistics of getting the lockdown (isolation) center established and operationalized.
[10]TSL in Jamaica.

and implemented a revised ackee program that would eliminate the initial obligatory inspections by the FDA prior to a firm becoming green listed. Instead, for Jamaica, the BSJ would conduct a complete GMP and HACCP inspection and issue export certificates to companies only if they were in compliance with FDA requirements for ackee. For companies trying to get off DWPE in the future, the option of an isolation facility in-country prior to shipping is unlikely to be available. The process that has been agreed is that product (five separate lots) would first have to be shipped to a US port and then sampled for testing. The samples have to be sent to Jamaica for testing, followed by an analytical package being sent to the FDA's SRL for review and evaluation. Jamaican processors indicated that this is not a cost-effective option, but it remains the only option. It is also the route that has been used for the approval of ackees from other countries,[11] which also need to have their products meet the requirement of five shipments within the 100 ppm limit, and this testing is also carried out in Jamaica.

As part of the process of institutionalizing the gains of the collaboration over the years, the FDA's team upgraded an inspection visit in January 2012 under the current FSMA[12] inspection mandate to become a training/inspection trip. This was a reflection of the FDA's continued commitment to the partnership approach in developing food safety and security for trading partners with the United States for products falling under their remit. The FDA's approach to the ackee export trade has been highly appreciated by the BSJ, trade associations, and industry members in Jamaica. It has contributed in no small way to facilitating the realization that HACCP systems can be developed and implemented in a sustainable manner in their processing plants. This has led to a further wave of regulatory change in the country whereby HACCP-based food safety systems are becoming mandatory (Walters-Gregory, 2012), with the regulatory authorities in the Jamaican ministries of health, agriculture, and industry all promulgating HACCP-based regulations to manage food handling, trade, and export. Jamaica and, by extension, the whole Caribbean region will benefit from this process as it improves food safety management and thereby the competitiveness of the industry and its exports.

---

[11]Haitian ackee exports were approved in 2006 and Belizean ackee exports in 2009, the latter using this US-held shipment/Jamaican testing route for approval.
[12]The Food Safety Modernization Act, 2011.

CHAPTER 8

The Food Safety Modernization Act and Its Impact on the Caribbean's Approach to Export Market Access

Joyce Saltsman[1] and André Gordon[2]

[1]Retired from the Center for Food Safety and Applied Nutrition (CFSAN), US Food and Drug Administration, Baltimore, Maryland, USA
[2]Technological Solutions Limited, Kingston, Jamaica

ABSTRACT

The Food Safety Modernization Act, passed in the United States in 2011, was a major reform of previous US food safety laws. It involved a shift in focus from responding to food contamination events to prevention-based controls for food manufacture, harvesting, processing, packing, and storage. Exporters in developing countries need to

Food Safety and Quality Systems in Developing Countries. http://dx.doi.org/10.1016/B978-0-12-801227-7.00008-1

understand the principles of the Act and how it is implemented to avoid their products being rejected at US ports of entry. This chapter discusses the background to and major provisions of the Act, and the impact it has had on the Caribbean. In seeking to comply with the requirements of the Act, food-exporting companies and regulators in the region have begun a transformation of their food production, handling, and export processes. This is leading to improvements in their infrastructure and competitiveness, and will ultimately result in these companies gaining greater market access and market share globally.

Keywords: export-market access; Food Safety Modernization Act (FSMA); Global Food Safety Initiative (GFSI); good manufacturing practices (GMPs); preventive controls

8.1 INTRODUCTION

Food safety and quality systems in most developing countries have been driven either by the nascent and developing infrastructure in public health and consumer protection or by the need to meet the requirements in export markets. In both instances, consumers in both developing and developed countries have been the beneficiaries, as has also a less burdened health system. While local regulations have been an important driver, since the turn of the millennium the major driver for developing-country producers and exporters has been the ever-increasing requirements of their importing markets and their direct customers. Currently, the Global Food Safety Initiative (GFSI) set of equivalent benchmarked standards (Global Food Safety Initiative, 2014) are creating a private-sector-led revolution in the global food industry. Of equal and perhaps more far-reaching importance, certainly to firms intending to continue or commence exports to the United States, are the regulatory changes that have taken place in that country. These changes and their impact on selected developing countries in the Caribbean are the focus of this chapter.

The US Centers for Disease Control (CDC) estimates that each year one out of six people (approximately 48 million people) in the United States suffers from food-associated illnesses, resulting in more than 100,000 hospitalizations, and thousands of deaths (United States Food and Drug Administration (US FDA), 2014a). Although many food-borne illnesses result from improper food handling practices in the home and in food service settings, the agency primarily responsible for food safety, the US Food and Drug Administration (FDA), believes that the food industry can do a better job in its role to ensure the safety of processed foods and fresh fruits and vegetables. To address the prevailing food safety concerns, US law makers passed the Food Safety Modernization Act (FSMA) that required the FDA to drastically revise existing food safety regulations and develop additional regulations to address the global food market and associated food safety concerns. The FSMA is the most sweeping reform of food safety laws in the United States in more than 70 years. It shifts the focus from responding to food contamination events to preventing them.

The FSMA (Pub. L. 111-533), signed into law on January 4, 2011, represents one of the most aggressive efforts by the US government to address concerns about the burden of food-borne illnesses in the United States. The FSMA gives the FDA the authority to better protect public health by strengthening the safety and security of the food supply by updating current good manufacturing practices (cGMPs) under Volume 21, Part 110, of the Code of Federal Regulations (21 CFR 110). This focuses on prevention-based controls for how foods are manufactured, processed, grown and harvested, and packed and held, providing stronger enforcement authority to help achieve higher rates of compliance with FSMA regulations and better responses to problems when they occur. These controls are built on a foundation that relies on the best available science, science-based controls, and common-sense approaches to prevent problems that can make people sick. When fully implemented, FSMA regulations will affect most foods and food ingredients in the farm-to-table food chain, including those exported to the United States. For this reason, exporters from developed countries, including Latin America and the Caribbean, need to understand the principles upon which the Act is based and the details of how it is being, and will be, implemented as it rolls out each successive provision (Gordon, 2013a).

8.2 MARKET-ACCESS ISSUES FOR CARIBBEAN EXPORTS TO THE UNITED STATES

The United States remains an important market for exports from Latin America and the Caribbean (LAC), although its share in the region's trade had dropped from 60% in 2000 to 39% by 2009

(ECLAC, 2011b). Of these exports, food and agricultural exports are a major component. It is therefore important that countries of the region are able to access the US market with food exports without facing the problem of rejection at the ports of entry. The percentage of foods being rejected at US ports of entry from the Caribbean both prior (Figure 8.1) and subsequent to the implementation of the FSMA (Table 8.1) remains very low. However, at 17%, the Caribbean was second only to Mexico in the number of items rejected and in 2004, 2006, and 2007 had more foods rejected at US ports of entry than were accepted into the country (Figure 8.1). Further, the importance of the rejected items varies in its impact on each country's agrifood export sector. It is imperative, therefore, that Caribbean countries understand the reasons for the failure of their food exports to comply with US import requirements both before and after the implementation of the FSMA if they are to maintain and grow their exports to that market.

A breakdown of the food products from the Caribbean refused at US borders between 2002 and 2010 shows that the major product categories that faced challenges were foods of vegetal origin, other processed foods and fruits, and related products (Figure 8.2). Rejections of other processed foods, beverages, and fruits and subproducts were higher in the period 2006–2010 than in the prior period. The main reasons for rejections were the presence of pesticides, filth, labeling issues, and missing information (Figure 8.3). Producers and exporters from Latin America and the Caribbean, even before the advent of the FSMA, had deficiencies in their production and food handling systems that resulted in refusals at the US border. These required better processing practices to correct them and also a good understanding of the applicable standards, laws, and regulations with which they needed to comply, as well as industry best practices to ensure that the exports would consistently meet the requirements. The FSMA, with its mandate for increased inspection at ports of entry, as well as increased scrutiny of importers and exporters, will result in continued and even higher levels of rejection if exporting developing countries do not quickly get up to speed, as the United States seeks to ensure a safer, more compliant food supply for its people.

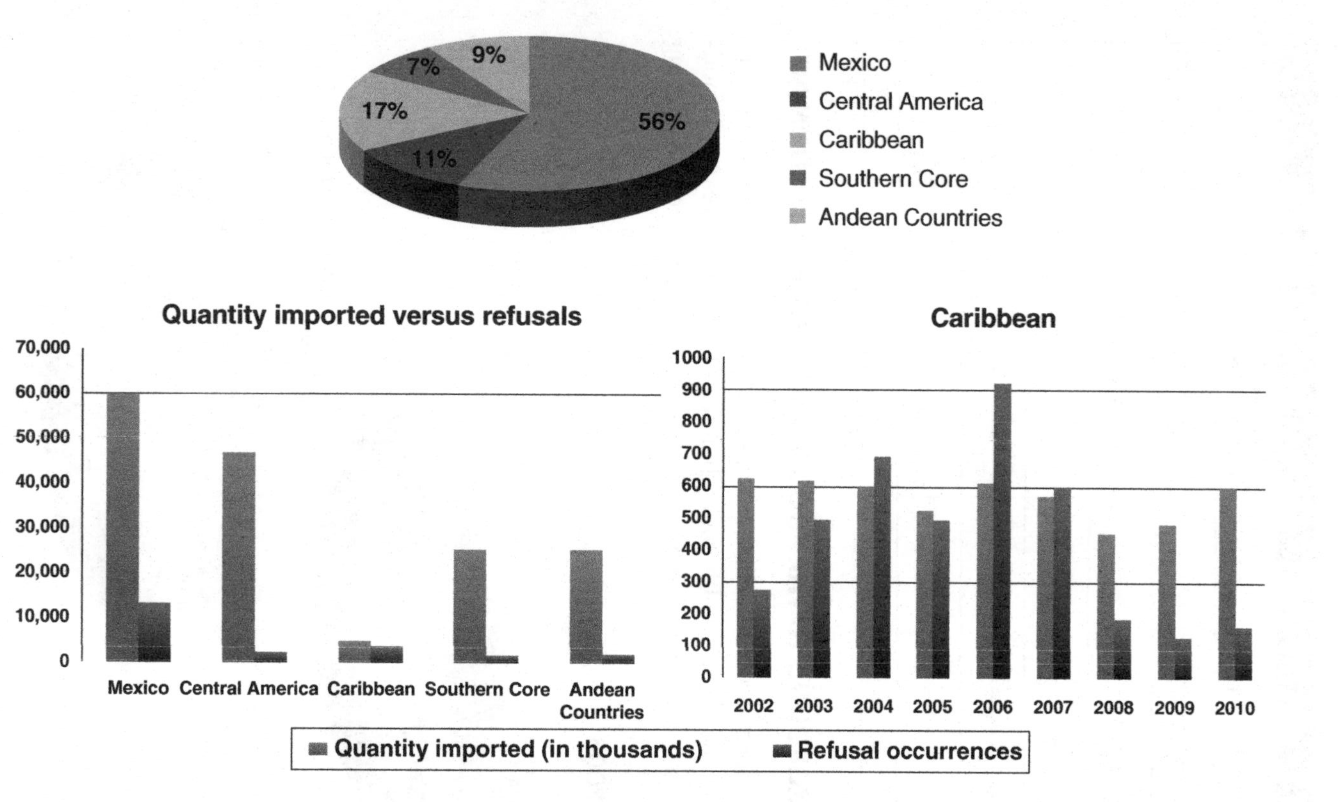

Fig. 8.1. Refusals of agricultural products from LAC at US borders, 2002–2010. (Source: ECLAC, 2011a.)

Table 8.1. US Rejections of Imported Food by Country, 2012

Country	Food Imports Rejected	No. of Product Lines Imported into the United States	Rejection (%)
Dominican Republic	442	907,927	0.05
Puerto Rico	6	249,015	0.00
Jamaica	92	17,906	0.51
Trinidad and Tobago	40	11,060	0.36
Guyana	80	5600	1.43
Suriname	39	2179	1.79
Barbados	14	1812	0.77
Haiti	27	1501	1.80
Grenada	9	1419	0.63
Bahamas	3	1214	0.25

Source: Gordon, 2013b.

8.3 BACKGROUND TO THE FOOD SAFETY MODERNIZATION ACT

It has long been recognized that a breakdown in food safety standards at any point in the farm-to-table food chain can lead to outbreaks of food-borne illnesses, putting a significant burden on public health and causing

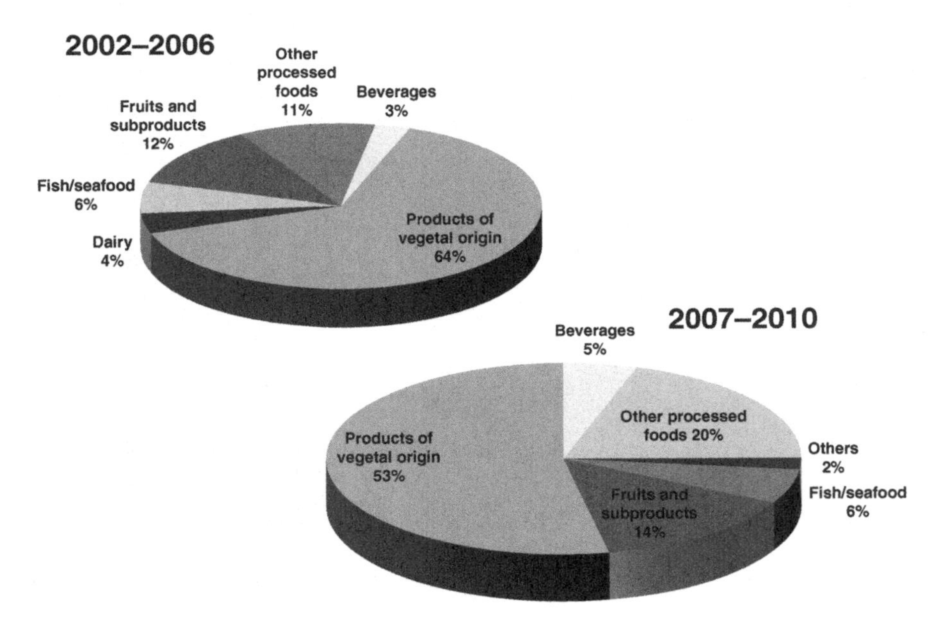

Fig. 8.2. Products refused entry at US ports, 2002–2010. (Includes Caribbean Community – CARICOM members, Cuba, and Dominican Republic.) (Source: ECLAC, 2011a.)

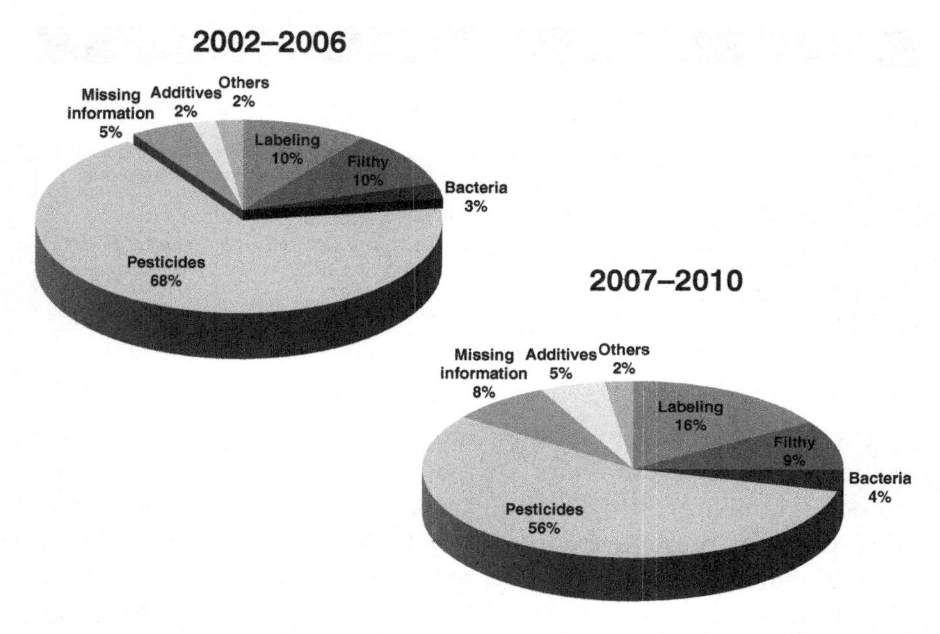

Fig. 8.3. Problems found in Caribbean agriproducts at US ports, 2002–2010. (Includes Caribbean Community – CARICOM members, Cuba, and Dominican Republic.) (Source: ECLAC, 2011a.)

great disruption and economic loss to the food industry (Hamburg and Sharfstein, 2009). When looking at the food supply chain as a whole, it becomes evident that the FSMA puts the responsibility for the safety of foods introduced into the US food supply system squarely on the shoulders of those involved in its production, processing, packaging, holding, and transport. Under the FSMA, food processors must evaluate their processing operations and identify hazards that could contaminate the food, implement and monitor effective measures (controls) to prevent such contamination, and identify the situation and take corrective measures should contamination occur. Similarly, farms must evaluate their production, harvesting, packing, and holding practices to ensure that they comply with the science-based standards established by regulation for fruits and vegetables consumed without further processing, such as cooking. There are, however, exemptions to both the preventive control for human foods regulation and the Produce Safety regulations for very small businesses and farms (US FDA, 2014b,c), which will also apply to some exporters from developing countries. It is the responsibility of these exporters and their importers to determine whether they qualify for the exemption (Table 8.2).

Table 8.2. Elements of the Food Safety Modernization Act, 2011

- Revised cGMP requirements
- Hazard analysis and risk-based preventive controls for human food (Preventive Controls for Human Food)
- Hazard analysis and risk-based preventive controls for animal food (Preventive Controls for Animal Food)
- Standards for growing, harvesting, packing, and holding of produce for human consumption (Produce Safety Standards)
- Standards for importers (Foreign Supplier Verification Program)
- Program for accrediting public and private bodies to provide credible certifications that regulated entities are meeting US safety standards
- Program for accrediting laboratories
- Sanitary transportation of food (human or animal) standards

The United States imports an estimated 15% of its food supply, including 50% of fresh fruits, 20% of fresh vegetables, and 80% of seafood (Hamburg, 2014). It is no surprise, therefore, that the FSMA regulations for preventive controls for human food and the produce safety requirements apply to imported foods in much the same way as they do to US-produced foods. What differs is the approach. Under the authority given to the FDA under the FSMA, the agency stepped up its authority to oversee the millions of products imported into the country annually and to require importers to verify the safety of food from suppliers. Not only has the FDA stepped up its review of foods presented for import at the borders; it has also increased the numbers of foreign facility inspections, an important part of the imported food safety requirements. This was evident in the significant increase in Caribbean inspections in 2012–2013 (Gordon, 2013a). Should the exporting facility or foreign government refuse access to food facilities, the FDA may block foods from either the specific facility or the country from coming into the United States.

8.4 MAJOR PROVISIONS OF THE PREVENTIVE CONTROLS FOR HUMAN FOOD FINAL REGULATION

The final regulation on preventive controls for human food applies to facilities that manufacture, process, pack, or hold human food and that are registered, with some exceptions, with the FDA under the current food facility registration regulations (US FDA, 2014d). To be compliant with the preventive control rule, processors must establish the following:

- A written food safety plan – a plan that addresses the safety of foods throughout manufacturing, processing, packing, or holding operations at each facility.
- Hazard analysis – identification and evaluation of known or reasonably foreseeable hazards for each type of food manufactured, processed, packed, or held at a facility.
- Risk-based preventive controls – identification and implementation of control measures to provide assurance that the hazards that are reasonably likely to occur will be significantly minimized or prevented. Preventive controls would not be required for those facilities that determine that there are no hazards reasonably likely to occur. Preventive controls are required to include, as appropriate: (1) process controls, (2) food allergen controls, (3) sanitation controls, and (4) a recall plan.[1]
- Monitoring procedures – providing assurance that the established preventive controls are consistently performed.
- Corrective actions – actions required when preventive controls are not properly implemented. The FDA requires the facility to evaluate the food for safety and prevent affected food from entering commerce where necessary. In the absence of specific corrective action procedures, or if the preventive control fails to be effective, the operator should re-evaluate the food safety plan so modifications can be made.
- Verification – activities to ensure that preventive controls are consistently implemented and effective. Several approaches may be considered for this activity. For example, validation could be used to determine if preventive controls are adequate and effective for their intended purpose, activities undertaken to verify that controls applied to a process are operating as intended, and a review of monitoring records carried out.
- Recordkeeping – records associated with food operations, such as the written food safety plan, records of preventive controls, monitoring activities, corrective actions, and verification activities.

Hazard analysis and risk-based preventive control requirements are similar to hazard analysis critical control points (HACCP) systems,

[1]The FDA determined that preventive controls for raw materials and ingredients can be covered under a supplier approval and verification program.

which the FDA, heretofore, has required for juice and seafood products. Operators of a manufacturing or processing facility, with some exceptions, are required to understand the hazards that are reasonably likely to occur in their operation and to implement preventive controls to minimize or prevent the hazards. An important difference between a hazard analysis and preventive control approach and HACCP is that preventive controls may be required at points other than critical control points and critical limits would not be required for all preventive controls (US FDA, 2014b).

8.5 MAJOR PROVISIONS OF THE STANDARDS FOR GROWING, HARVESTING, PACKING, AND HOLDING OF PRODUCE FOR HUMAN CONSUMPTION (PRODUCE SAFETY RULE)

The Produce Safety rule focuses on microbiological hazards related to growing, harvesting, packing, and holding fresh produce. The rule covers all fruits and vegetables except those rarely consumed raw, produced for personal consumption, or destined for commercial processing that will reduce the number of microorganisms of public health concern. The final rule is based on science and risk-analysis, and therefore focuses on areas of risk, most noticeably:

- Worker health and hygiene – requires all personnel, including supervisors, who handle fresh produce or food-contact surfaces to be qualified and trained, and establishes hygienic practices and other measures needed to prevent personnel, including visitors, from contaminating produce.
- Agricultural water – requires water used for agricultural purposes to be:
  - Safe and sanitary for its intended use, including water that may contact the harvestable portion of the food or food-contact surfaces
  - Treated, if it is not naturally safe, and of adequate sanitary quality for its intended use
  - Periodically tested and action taken when water does not meet quality standards specified in the final rule
- Biological soil amendments – the rule provides requirements for determining whether the soil amendment has been treated or not; it prohibits the use of human waste, except when in compliance with

the Environmental Protection Agency (EPA) regulations for produce covered by the regulation; and establishes application requirements and minimum application intervals.
- Domesticated and wild animals – establishes standards for when animals are allowed to graze or be used for work in production fields, and requires monitoring of areas prone to animal intrusion prior to harvest and, as needed, during the growing season.
- Equipment, tools, and buildings – establishes requirements for equipment and tools used with fresh produce and instruments and controls, buildings, domesticated animals in and around fully enclosed buildings, pest control, hand-washing, and toilet facilities, sewage, trash, plumbing, and animal excreta.
- Recordkeeping – requires certain records to be kept.
- Sprout production – establishes measures to be taken to protect seeds or beans from contamination during growing, harvest, packing, and holding of sprouts.

The Produce Safety rule is the result of extensive outreach by the FDA with consumers, government, industry, researchers, and many others. It builds on existing voluntary industry guidelines for food safety that many producers, growers, and others currently follow. The rule is aimed at being flexible for different-sized farms, at complementing conservation laws and rules, and at not conflicting with laws and rules for organic farming. Exemptions to the Produce Safety rule are based on average sales of produce over a 3-year period (US FDA, 2014c).

8.6 MAJOR PROVISIONS OF THE FOREIGN SUPPLIER VERIFICATION PROGRAMS FOR IMPORTERS OF FOOD FOR HUMANS OR ANIMALS

The Foreign Supplier Verification Programs (FSVP) require importers to create and follow a risk-based program to help ensure the safety of the foods they import. Although the requirements vary depending on the type of food imported (e.g., fresh produce, processed foods, or dietary supplements), and the category of food (discussed further), importers must ensure the foreign supplier's foods intended for export to the United States are in compliance with processes and procedures, including risk-based preventive controls, to provide the same level of public health protection as that required for domestically produced foods. The FSVP

Table 8.3. Foods Exempt from the Foreign Supplier Verification Programs Regulations

- Juice products subject to FDA HACCP regulations
- Fish and fishery products also subject FDA HACCP regulations
- Foods for personal consumption
- Alcoholic beverages
- Transshipped foods (imported foods intended for export)
- Foods used for research or evaluation

Source: US FDA, 2014e – FSVP final rule.

Table 8.4. Foods Subject to Modified Requirements under the Foreign Supplier Verification Programs

- Low-acid canned foods
- Dietary supplements
- Foods from very small importers or imports from a very small supplier (<US$500,000 in annual food sales)
- Imported food from a supplier in a country with an officially recognized or equivalent food safety system

Source: US FDA, 2014e – FSVP final rule.

final rule applies to most foods under FDA jurisdiction, while exempting certain categories of food or modifying requirements for certain foods and suppliers, as shown in Tables 8.3 and 8.4.

The FSVP final regulation requires importers to maintain a written list of their foreign suppliers and establish written verification procedures in order to verify that hazards identified as reasonably likely to occur in a food being imported are being adequately controlled. If the importer or its customer is controlling a hazard, the importer must document such control. For other hazards, there are two options for verification activities. Option 1, on-site auditing of the foreign supplier, would be required for hazards to be controlled by the foreign supplier when there is a reasonable probability that exposure to the hazard will result in serious adverse health consequences or death. For certain raw fruits and vegetables that are typically consumed raw, auditing of the exporters' facility would be required for microbiological hazards. For other hazards, including less serious hazards and those that the foreign supplier has verified to be controlled by its supplier, importers may choose, under option 1, verification activities or activities that will provide sufficient assurance that the hazards are controlled. For example, such activities might be on-site auditing of the foreign supplier, periodic lot-by-lot sampling and testing, and periodic review of the supplier's food safety records. Other activities could also be considered if appropriate to verify the hazard is being controlled adequately.

Table 8.5. Key US Regulations Affecting Developing-Country Food Exports

Preexisting Regulation	New Regulations Under the FSMA
21 CFR 101 – Food Labeling	Right to Administrative Detention
21 CFR 110 – Current Good Manufacturing Practices (cGMPs)	Preventive Controls for Human Food
21 CFR 113 – Low-Acid Canned Foods	Produce Safety Rule
21 CFR 114 – Acidified Foods	Sanitary Transportation of Foods
21 CFR 120 – Hazard Analysis Critical Control Points (Juice)	FSMA Traceability and Recall Requirements
21 CFR 123 – Hazard Analysis Critical Control Points (Seafood)	Voluntary Qualified Importer Program
Nutrition Labeling and Education Act – Nutritional Labeling	Foreign Supplier Verification Programs
Food Allergen Labeling and Consumer Protection Act – Allergen Labeling	FSMA Allergen Management Requirements
21 CFR 11 – Electronic Records; Electronic Documents	Accreditation of Third-Party Auditors
21 CFR 75 and 82 – Labeling of Certified Colors	Laboratory Accreditation

Source: Gordon, 2013b.

Option 2 allows importers to choose verification activities for all types of hazards not controlled by the importer or its customer (US FDA, 2014e). Under this option, the frequency of verification activities would be determined by the nature of the risk presented by the hazard, the probability of the hazard causing serious health problems, and the food's and foreign supplier's compliance with US food safety regulations. Other requirements for the importer under the FSVP regulation include reviewing complaints, investigating adulteration or misbranding (e.g., in cases of allergen labeling), and taking corrective actions when the supplier is not compliant with the program. The importer's FSVP must be reviewed every 3 years or whenever the importer becomes aware of new information about potential hazards associated with the food being imported.

These regulations, along with the others identified in Table 8.5, will impact food exporters.

8.7 FDA INSPECTIONS IN THE CARIBBEAN AND THE FSMA

The FDA has been conducting relatively infrequent periodic inspections of food handling facilities in the Caribbean over the last 12 years, mainly of seafood plants exporting to the United States, which are under the jurisdiction of its seafood HACCP regulation (21 CFR 123), which was

Table 8.6. Inspections Conducted by the FDA in Jamaica and in Trinidad and Tobago in 2011–2013

Country	Period	Number of FDA Inspections	Facility Inspected
Jamaica	2011	1	Packing house
	2012	14	3 Packing houses; 11 processed food facilities
	2013	26	Processed food facilities
Trinidad and Tobago	2013	16	4 Low-acid canned food/acidified food, 6 snack foods, and 6 seafood plants

issued in 1995. These inspections sought to assure the agency of the compliance of Caribbean exporters with its requirements. Also, from 1999 to 2008, the FDA conducted several inspections of nine low-acid and acidified food plants exporting canned ackee and other products to the United States (Gordon, 2013b). This was part of a program focused mainly on ackee (discussed in Chapters 6 and 7). Starting in 2011 after the passage of the FSMA, but mainly during the period January 2012 to December 2013, the FDA undertook a significantly increased program of inspections in the Caribbean, with Jamaica and Trinidad and Tobago being the major focus among the English-speaking countries (Table 8.6).

The inspections in Jamaica and in Trinidad and Tobago were conducted by relatively experienced inspectors who, covering the expected areas, were mainly focused on cGMPs and other regulations in force at the time of inspection. The level and depth of the inspections were dependent on the inspector, their experience, and related factors (Gordon, 2013a). The first round in Jamaica started in 2011 and focused mainly on produce and low-acid canned food (LACF)/acidified food (AF) exporters (Table 8.6), all of which were found to be compliant with requirements. The second round of inspections, after the FSMA was passed, was from May to August 2012, covered a range of different types of exporters, and saw the issuing of FDA Form 483 to 12 firms, with observations to drive corrective actions, where applicable. One firm was issued with a warning letter (Gordon, 2013b). An FDA Form 483 is issued to the management of a firm at the end of an inspection when an investigator (inspector) has, in their judgment, observed any conditions that may constitute violations of the Food Drug and Cosmetic (FD&C) Act and related acts (US FDA, 2014f). Each observation noted on the FDA Form 483 is typically clear, specific, and significant. The observations recorded by the investigator indicate conditions or practices observed that have either adulterated the product or indicate that the product is being prepared, packed, or held under conditions that could cause it to be adulterated or otherwise become injurious to health of consumers (US FDA, 2014f). A warning letter is a formal correspondence issued to the principals of a firm about matters that have been subject to prior discussion and notification by the FDA but on which no or inadequate action has been taken by the firm. The matter raised with the Jamaican firm has since been resolved.

In 2013, the FDA conducted 26 inspections in Jamaica (bakeries, dairy, other products) and 16 in Trinidad and Tobago (LACF/AF, snack foods, seafood, bakeries, others). By this time, the traceability rule under the FSMA was in force and traceability, management of allergens, and labeling were among the areas examined, along with compliance with 21 CFR 110 (cGMPs), 21 CFR 113 (LACF), and 21 CFR 114 (AF), among others. As an outcome of the inspections in Jamaica, 12 FDA Form 483 VAIs (voluntary action indicated) were issued, with six firms being fully compliant (no 483s issued). In Trinidad and Tobago, about half of the firms were issued 483s, with the others, mostly seafood exporters being inspected under 21 CFR 123 and a confectionary and snack food firm, being fully compliant.

8.8 IMPACT OF THE FSMA ON THE CARIBBEAN AND THE APPROACH TO EXPORT MARKET ACCESS

The implementation of the FSMA, while requiring that exporters to and participants in the food industry in the United States meet the requirements of the Act, should not have come as a surprise to the food industry. Between 2002 and 2006, through the establishment of the European Food Safety Authority (EFSA) and the passage of several critical directives and regulations, the EU had moved to a HACCP-based system of ensuring the safety of its food supply. Since 2007 the major brands, major retailers, and distributors in the global food industry have been using the GFSI set of standards as prerequisites for doing business, and the Canadians have subsequently passed the Safe Food for Canadians Act, 2012, to be implemented in 2015 (Canadian Food Inspection Agency, 2012). The requirements of the FSMA will prove challenging for some producers and exporters from developing countries, as they will to some US-based producers. However, they also provide an opportunity for upgrading operations to word-class levels of competitiveness and thereby gaining greater market access and market share.

The impact of the changes in regulatory and market requirements, including FDA inspections and the FSMA, on regulatory practices in the Caribbean has been significant. The reality of food-safety-driven external inspections has resulted in a focus on the realignment of regional regulatory practices to become more food safety-driven. Throughout the region, in Barbados, Jamaica, Suriname, the Dominican Republic, and Trinidad and Tobago, as well as in Grenada, St. Lucia, and Dominica (members of the Organization of Eastern Caribbean States – OECS), the ministries of agriculture, health, and commerce and trade have increased their vigilance of the domestic production, handling, and sale of food based on food safety principles (Gordon, 2013a). Most of these countries have developed and implemented programs to sensitize the industry to the changing requirements and support them in making the transition to science-based food safety-driven food handling systems, rather than those based primarily on quality. Of these countries, Jamaica has advanced furthest down this path, largely owing to its interaction with the FDA and the series of inspections that started in 1999. These, buttressed by EU inspections of the seafood export sector in the early 2000s, have resulted in significant transformation of Jamaica's regulatory infrastructure. The enactment of a series of regulations encompassing agriculture, health, commerce, and export trade now sees a much greater focus on food safety rather than regulation-determined, standard-of-identity-driven quality requirements. HACCP-based food safety systems, with good manufacturing practices (GMPs) as their base, are slowly becoming a basic requirement for entry into or continuance in the food-handling sector. This will improve compliance of exporters with import market requirements in the EU, Canada, and the United States, with compliance with the FSMA being both a major driver and a major focus of regional governments.

The advent of the FSMA and the resultant focus on ensuring compliance with market access and regulatory requirements are resulting in a significant transformation in production and food handling practices of exporters from the Caribbean region. These changes are taking place at the farm level and throughout the food supply chain. This has involved regulatory, voluntary, internationally funded, and private sector-driven programs to implement science-based good agricultural practices (GAPs), GMPs, validations of process and production systems, and HACCP-based controls in the food production and handling sector. The implementation of traceability and recall programs, better-trained staff, and greater flexibility and room for growth are direct results of this thrust. It has also led to other tangible benefits for the sector in greater production efficiencies, lower losses, significantly improved food safety, the use of validated production processes, and the development of scientific bases to support traditional production processes. It has also resulted in better replicability and scalability of production processes because of the more structured nature of production and handling processes and, critically, greater export growth.

In general, the thrust towards science-based food safety controls and compliance with the FDA's and the FSMA's requirements has brought business benefits to the food industry of the region as a whole. This should be replicated across other developing countries as they seek to access the benefits that can be derived from doing business in the world's largest market.

8.9 CONCLUSIONS

The passing of the FDA's Food Safety Modernization Act, 2011, and its phased implementation have necessitated that exporting firms understand its requirements and seeking to comply with the preventive controls, produce safety, and foreign supplier verification programs, among other requirements. It has started a transformation in food production, handling, and export in the Caribbean region, enhancing steps already being taken in this direction in countries such as Jamaica, who already had an ongoing relationship with the FDA. While it has created and will create challenges for compliance of many firms involved in food handling and exports, the FSMA will also create opportunities for many. It has been a catalyst for firms to undertake upgrading operations to word-class levels of competitiveness, including improving physical infrastructure, enhancing their human-resource base, and implementing a range of productivity-improving systems. This will ultimately result in the firms that make the grade gaining greater market access and market share in the United States and other markets, thereby enhancing their sustainable profitability and the livelihood of their staff.

APPENDIX I

Important Events in the History of the Export of Jamaican Ackees to the United States of America

Year	Action	Outcome
1973	FDA Import Alert 21–11 imposed on canned ackee (*Blighia sapida*) because of insufficient evidence of its freedom from HGA and HGB. HGA classified as a non-GRAS food additive	Legal exports of canned ackees and ackee products to the USA prohibited. Firms automatically on "Detention Without Physical Examination" (DWPE) which means any product arriving at a US port was automatically denied entry without any inspection of the product and had to be shipped elsewhere or destroyed
1978	The BSJ (JBS) asks the FDA to supply the exact protocol for the toxicological studies required to prove the safety of HGA	The FDA indicated that they had not developed any protocol but would accept guidelines from the National Academy of Sciences – National Research Council
1979	In an effort to supply the information required by the FDA, the BSJ developed a project consisting of three phases: 1. The development of a reliable method for the chemical analysis of HGA in ackee aril and brine 2. A toxicological evaluation of the ackee involving feeding studies using three animal species 3. The development of processing techniques designed to produce a toxin-free product. Assistance was sought from the World Bank	Mr. Gladstone Rose, a UWI Graduate in Special Chemistry, working in the chemistry laboratory, began a review to compare the existing methods of hypoglycin A analyses in ackee to determine suitability for the routine monitoring of the toxin in canned ackees
1983	The Chemistry Department of the BSJ received a high performance liquid chromatography (HPLC) technique through assistance from the World Bank	The BSJ now had the capability to do a much wider range of analyses than was previously possible
1983	Five methods for determining HGA were evaluated, including classical methods by Kean (1974). Gas chromatography and HPLC were also evaluated	Mr. Rose completed the study he had undertaken. He recommended HPLC as the approach with the greatest potential once a suitable method using it was refined for using it for HGA
1987	USAID funded technical assistance from Dr. Jorge Augustin of the Postharvest Institute for Perishables (PIP), University of Idaho (UI), and Dr. Robert Bates, University of Florida (UF), Gainesville. The assistance was to establish the feasibility of and mechanism for providing technical assistance to the Government of Jamaica's Jamaica Ackee Project (JAP)	A comprehensive report by Dr. Jorge Augustin of the PIP (UI) on the ackee issue. Dr. Bates, who visited subsequently, had discussions with the USAID, the JBS, the Food Technology Institute (FTI), and Dr. E. Kean of the Biochemistry Department, University of the West Indies and visited ackee processing plants. Dr. Bates and the UF, along with Ms. Andrea Robbins, BSJ project coordinator, developed and presented a proposal to USAID for a joint project, the JAP, aimed at developing a suitable analytical method, which would be used to provide supporting data for a petition to the USFDA for revising the Import Alert on Canned Ackees

(Continued)

Food Safety and Quality Systems in Developing Countries. http://dx.doi.org/10.1016/B978-0-12-801227-7.00016-0

Year	Action	Outcome
1988	Commencement of the JAP involving preparation of a verifiably pure HGA standard, validation of the analytical method and studies on the commercial product	Successful extraction and purification of HGA by the JBS Chemistry Department based on a modification of a method developed by Kean (1974)
1989	Work on a validated method for the determination of HGA concluded at the UF under the JAP	Publication of the method in the McGowan et al. (1989a) paper
1990	Chemical studies on commercially canned ackees undertaken to properly characterize the product	Ongoing application of the HPLC-based methodology developed by the UF team for HGA. First set of studies on commercially canned ackees completed and information compiled for the JAP final report
1990	Chemical studies completed by Ms. Ingrid Ashman of the BSJ Chemistry Department, including analytical data for commercial canned ackee	Data on HGA in ackees from Jamaican factories obtained. Final JBS project report for the JAP submitted to USAID by Ms. Andrea Robins
1991	Presentation of the final project report from the UF and the BSJ	The final report from the UF recommended that: 1. The UF and JBS should continue analysis and data collection 2. The JBS should continue working with the industry to establish industry norms under controlled but realistic conditions 3. The JBS should work with the industry to establish the feasibility of producing hypoglycin free ackee as well as defining the lowest practical level which could be achieved Other recommendations included: a. Establishing dialog with the FDA to review the data collected as well as revising the need for toxicological studies b. Developmental work on the analytical procedure developed by the UF
1991/1992	Studies on ackees, including canned ackees and ackee ripening, completed at the UF	a. Establishment/formalization of a maturity index for ackees b. Documentation of the levels of HGA associated with each stage of maturity c. Establishment of likely levels of HGA in canned ackees d. Publication of milestone research by the UF team in Journal of Food Safety, Vol. 12 (2) (1991/1992)
1992	Ongoing monitoring of the hypoglycin levels of Jamaican canned ackees by the BSJ	Database of HGA results on canned ackee from across the island established
1992	Ongoing monitoring of HGA levels in canned ackees commenced in earnest	Database of HGA results on canned ackee substantially being expanded by new data
1993	Visit to the FDA by Jamaican delegation to review the Import Alert[1]	FDA was interested in having an evaluation done of the analytical method by IUPAC and indicated that more detailed evidence was required to establish the safety of HGA as a food additive

(Continued)

Year	Action	Outcome
1994	Health Canada requested hypoglycin A standard material to assist them in the monitoring of ackee imports	The sample was supplied and this facilitated Health Canada in establishing their limit for HGA
1995–1997	Continuation of the monitoring program for hypoglycin in canned ackee	Ms. Olive Grossett, BSJ, working with Mr. James Kerr, continued to expand and compile an extensive body of data on canned ackee
October, 1998	Jamaica Ackee Task Force (JATF) under the auspices of the JEA[2] and supported by the US Embassy through the USDA is established	A comprehensive program to transform the industry to make selected processors hazard analysis and critical control point (HACCP)-compliant and regulatory oversight HACCP-based is developed
April, 1999	Jamaican delegation led by the JEA visits the FDA and presents a comprehensive, science-based petition for the lifting of the Import Alert[3]	The FDA agrees to a program which, if successful, will allow the removal for the Import Alert for qualified exporters
April/May 1999	FDA sets up Jamaica Ackee Outreach Program[4]	Key collaborations are initiated. Objectives and goals are set for Jamaica and the FDA to work collaboratively foward for the first time
October/ November, 1999	The first FDA inspection team visits Jamaica to inspect plants and visit the JBS	A relationship between the industry and regulators in Jamaica and the FDA, supported by the US Embassy, is established
July 6, 2000	The Import Alert on Ackees Canned in Brine is lifted	Canco Limited and Ashman Food Processors Limited are able to commence exports to the USA

[1]The delegation consisted of the late Dr. Juliette Newell of Tijule Ltd. (representing the processors), Dr. George Wilson from the Jamaica Agricultural Development Foundation, and Mr. James Kerr, JBS and was facilitated by the FDA and supported by the US Embassy, the Jamaican Embassy, USDA, and USAID.

[2]The JATF was chaired by Dr. André Gordon, Managing Director of Technological Solutions Limited, a Jamaica Exporters Association (JEA) Director and included representatives of the Ministries of Industry and Commerce and Agriculture, the JBS, processors and distributors, funding agencies, and the USDA. A detailed composition of the Task Force in is Appendix II.

[3]This delegation included the Jamaican Ambassador, Dr. Richard Bernal, the USDA, the JEA, Processors, and the BSJ. The technical and industry presentation was led by Dr. Gordon and included Mr. James Kerr (JBS) and Mr. Norman McDonald (Canco Ltd.).

[4]The Jamaica Ackee Outreach Program was led by Dr. Joyce Saltsman, Interdisciplinary Scientist, Office of Plants, Dairy Foods, and Beverages (now the Office of Food Safety).

APPENDIX II

Key Participants in the Collaborative Program on Ackees from 1999 to 2012

FDA

- CFSAN – Office of Plants, Dairy Foods and Beverages (now the Office of Food Safety (OFS)): Dr. Joyce Saltsman, Project Lead
- CFSAN – Office of Compliance
- FDA/ORA: Mrs. Debra DeVlieger, Field Lead
- FDA/DIOP
- FDA Southeast Regional Laboratory (Mr. George Ware (retired))

US EMBASSY JAMAICA

- USDA Agriculture Attaché, Mrs. Yvette Perez

USDA

- Agricultural Statistician – NC State University/Agriculture Research Center (Dr. Thomas Whitaker)

JAMAICA

- Jamaican Export Association: Mrs. Pauline Gray, Mr. Hernal Hamilton, Mrs. Beverly Morgan, Dr. André Gordon (Chair, Jamaica Ackee Task Force – JATF)
- Bureau of Standards Jamaica (Jamaican Competent Authority): Mr. James Kerr
- Jamaica Agro-Processors Association (Mr. Michael Ming, Mrs. Dorothy Ramsey, Ms. Donna Bromfield)
- Technological Solutions Limited (Food Safety Specialist, Thermal Process Authority, Testing Laboratory for Hypoglycin A): Dr. André Gordon – Technical Lead, Ms. Veronica Morgan
- Science Research Council (Thermal Process Authority): Mr. Maurice Lewis

Food Safety and Quality Systems in Developing Countries. http://dx.doi.org/10.1016/B978-0-12-801227-7.00017-2

- Private Firms: Mr. John Mahfood (Grace Kennedy & Co. Ltd.), Mr. Norman McDonald (Canco Ltd.), Dr. Juliette Newell (Tijule Ltd.), Mr. Ira Ashman (Ashman's Food Products Ltd.), Mr. Andrew Morales (West Best Ltd.)

Also contributing were:

- The National Development Bank of Jamaica (Mr. Byron Wynter, Ms. Claudia James)
- Jamaican Rural Agriculture Development Authority/Ministry of Agriculture: Mr. Aaron Parke, Permanent Secretary
- Jamaica Promotions Corporation (JAMPRO): Mr. Robert Kerr
- Mona Institute of Applied Sciences: the University of the West Indies (provided analytical testing after 2008)

REFERENCES

Addae, J.I., Melville, G.N., 1988. A re-examination of the mechanism of ackee-induced vomiting sickness. West Indian Med. J. 37 (1), 6–8.

Al-Bassam, S.S., Sherratt, H.S.A., 1981. The antagonism of the toxicity of hypoglycin by glycine. Biochem. Pharmacol. 30 (20), 2817–2824.

Animal and Plant Health Inspection Service (APHIS), 2014. Proposed rule on the importation of kiwi fruit from Chile into the United States. Extracted on January 25, 2015. Available from: https://www.federalregister.gov/articles/2014/10/16/2014-24631/importation-of-kiwi-from-chile-into-the-united-states.

Ashurst, P.R., 1971. The toxic substance of ackee – a review. J. Sci. Res. Counc. 2 (1), 4–16.

Baldwin, J.E., Parker, D.W., 1987. Stereospecific (methylenecyclopropyl) acetyl-CoA inactivation of general acyl-CoA dehydrogenase from pig kidney. J. Org. Chem. 52 (8), 1475–1477.

Baldwin, J.E., Adlington, R.M., Bebbington, D., Russell, A.T., 1994. Asymmetric total synthesis of the individual diastereoisomers of hypoglycin A. Tetrahedron 50 (41), 12015–12028.

Baldwin, J.E., Ostrander, R.L., Simon, C.D., Widdison, W.C., 1990. Stereospecific inactivation of the general acyl-CoA dehydrogenase from pig kidney by (R)-(-)-(methylenecyclopropyl) acetyl-CoA and (S)-(+)-(methylenecyclopropyl) acetyl-CoA. J. Am. Chem. Soc. 112 (5), 2021–2022.

Barennes, H., Valea, I., Boudat, A.M., Idle, J.R., Nagot, N., 2004. Early glucose and methylene blue are effective against unripe ackee apple (*Blighia sapida*) poisoning in mice. Food Chem. Toxicol. 42 (5), 809–815.

Bates, R.P., 1991. Ackee project final report. Jamaica Bureau of Standards/United States Agency for International Development, University of Florida, Gainesville.

Berg, J.M., Tymoczko, J.L., Stryer, L., 2012. Biochemistry, seventh ed. W.H. Freeman, New York.

BIBRA, 1995. An Evaluation of the Toxicity of Hypoglycin A in Canned Ackees. BIBRA International, Carshalton, Surrey, United Kingdom.

Billington, D., Osmundsen, H., Sherratt, H.S.A., 1978. Mechanisms of the metabolic disturbances caused by hypoglycin and by pent-4-enoic acid in vitro studies. Biochem. Pharmacol. 27 (24), 2879–2890.

Blake, O.A., Bennink, M.R., Jackson, J.C., 2006. Ackee (*Blighia sapida*) hypoglycin A toxicity: dose response assessment in laboratory rats. Food Chem. Toxicol. 44, 207–213.

Blake, O.A., Jackson, J.C., Jackson, M.A., Gordon, C.L.A., 2004. Assessment of dietary exposure to the natural toxin hypoglycin in ackee (*Blighia sapida*) by Jamaican consumers. Food Res. Int. 37, 833–838.

Bliss, R.M., 2008. Jamaican delicacy makes a comeback – statistical research helped put ackee fruit back on U.S. grocery shelves. Agric. Res. 56 (5–6), 18.

Bowen, C.S., 2006. Evaluation of hypoglycins A and B content and the phytochemistry of *Blighia sapida*. PhD Thesis, Department of Chemistry, The University of the West Indies, Mona Campus, Kingston, Jamaica.

Bowen-Forbes, C.S., Minott, D.A., 2011. Tracking hypoglycins A and B over different maturity stages: implications for detoxification of ackee (*Blighia sapida* K.D. Koenig) fruits. J. Agric. Food Chem. 59 (8), 3869–3875.

Bowery, J.J., 1892. The Jamaica Gazette. 15, 99.

Food Safety and Quality Systems in Developing Countries. http://dx.doi.org/10.1016/B978-0-12-801227-7.00009-3

Bras, G., Jelliffe, D.B., Stuart, K.L., 1954. Veno-occlusive disease of liver with non-portal type of cirrhosis, occurring in Jamaica. Arch. Pathol. 57 (4), 285–300.

Bressler, R., 1976. The unripe ackee – forbidden fruit. N. Engl. J. Med. 295, 500–501.

Bressler, R., Corredor, C., Brendel, K., 1969. Hypoglycin and hypoglycin-like compounds. Pharmacol. Rev. 21 (2), 105–130.

Brooks, S.E.H., Audretsch, J.J., 1971. Hypoglycin toxicity in rats. II. Modification by riboflavin of mitochondrial changes in liver. Am. J. Pathol. 62 (2), 309–320.

Brooks, S.E.H., Miller, C.G., McKenzie, K., Audretsch, J.J., Bras, G., 1970. Acute veno-occlusive disease of the liver. Arch. Pathol. 89, 507–520.

Broughton, A., 1794. Hortus Eastensis. Spanish Town, Jamaica, p. 33.

Brown, M., 1989. Hypoglycin A: levels of maturing fruit of the ackee tree (*Blighia sapida*) and efforts towards its minimization in the canned product. M.S. Thesis, University of Florida, Gainesville.

Brown, M., Bates, R.P., McGowan, C., Cornell, J.A., 1992. Influence of fruit maturity on the hypoglycin A level in Ackee (*Blighia sapida*). J. Food Safety 12, 167–177.

Bureau of Standards Jamaica (BSJ), 1991. Jamaican Standard Specification for Processed Food (General) JS 36: 1991, Kingston, Jamaica.

Bureau of Standards Jamaica (BSJ), 2000. Jamaican Standard Specification for Canned Ackee (*Blighia sapida*) in Brine, JS 276: 2000 Kingston, Jamaica.

Bureau of Standards Jamaica (BSJ), 2012. The Jamaican Standard Specification for the Production of Processed Foods Utilizing the HACCP Principles (General) JS 317: 2012, Kingston, Jamaica.

Campbell, H., 2006. Ackee exports could resume next month. The Daily Gleaner, September 27, 2006, Kingston, Jamaica.

Canadian Food Inspection Agency, 2012. Natural toxins in fresh fruit and vegetables, consumer food safety fact sheets. Available from: http://www.inspection.gc.ca/food/information-for-consumers/fact-sheets/specific-products-and-risks/fruits-and-vegetables/natural-toxins/eng/1332276569292/1332276685336 (accessed 18.10.14.)

Caribbean Food and Nutrition Institute, 1998. Food Composition Tables for use in the English Speaking Caribbean, second ed. CFNI Press, Kingston, p. 37.

CentralAmericaData.com, 2012a. More fresh fruits on the table. In U.S. restaurants, tropical fruits are fast becoming favourites. Extracted on January 25, 2015. Available from: http://www.centralamericadata.com/en/article/home/More_Fresh_Tropical_Fruit_on_the_Table.

CentralAmericaData.com, 2012b. Increased consumption of mangoes in the U.S. Extracted on January 25, 2015. Available from: http://www.centralamericadata.com/en/article/home/Increased_Consumption_of_Mangos_inthe_US.

Centers for Disease Control (CDC), 1992. Toxic hypoglycaemic syndrome – Jamaica, 1989–1991. Morb. Mortal. Wkly. (MMWR) 41 (04), 53–55.

Chanel, M.I., 2001. Health-Haiti: ackee poisoning disturbs health officials. Inter Press Service News Agency, Port-au-Prince, Haiti. March 18, 2001. Extracted on November 5, 2014. Available from: http://www.ipsnews.net/2001/03/health-haiti-ackee-poisoning-disturbs-health-officials/.

Chase, W.C., Landen, W.O., Gelbaum, L.T., Soliman, A., 1988. Liquid chromatographic resolution of hypoglycin A from leucine. J. Chromatogr. 456, 431–434.

Chase, W.C., Landen, W.O., Gelbaum, L.T., Soliman, A., 1989. Ion-exchange chromatographic determination of hypoglycin A in canned ackee fruit. J. Assoc. Off. Anal. Chem. 72 (2), 374–377.

Chase, W.C., Landen, W.O., Gelbaum, L.T., Soliman, A., 1990. Hypoglycin A content in the arils, seeds and husk of ackee fruit. J. Assoc. Off. Anal. Chem. 73 (2), 318–319.

Crews, C., Clarke, D., 2014. Natural toxicants: naturally occurring toxins of plant origin. In: Mortajemi, Y. (Ed.), Encyclopedia of Food Safety, vol. 2. Hazards and Diseases. Elsevier Inc., pp. 261–268, Available from: www.sciencedirect.com/science/referenceworks/9780123786135.

DaSilva, C.A., Baker, D., Shepard, A.W., Jenane, C., Miranda-da-Cruz, S., 2009. Agro-Industries for Development. United Nations Food and Agricultural Organization/United Nations Industrial Development Organization, Rome, Italy.

Diop, N., Jaffee, S.M., 2005. Fruits and vegetables: global trade and competition in fresh and processed product markets. Global Agricultural Trade and Developing Countries. The World Bank, Washington, DC, pp. 237–257.

Dolan, L.C., Matulka, R.A., Burdock, G.A., 2010. Naturally occurring food toxins. Toxins 2 (9), 2289–2332.

Dundee, S.J., Minott, D.A., 2012. Impact of seed size on residual hypoglycin levels in ackee. Food Res. Int. 47 (2), 306–309.

ECLAC, 2011a. The United States and Latin America and the Caribbean. Data on economics and trade. The Economic Commission for Latin America and the Caribbean (ECLAC). United Nations. March 2011, Santiago, Chile.

ECLAC, 2011b. The United States and Latin America and the Caribbean. Highlights of economics and trade. The Economic Commission for Latin America and the Caribbean (ECLAC). United Nations. March 2011, Santiago, Chile.

Ellington, E.V., 1961a. A chemical contribution to the problem of vomiting sickness. Bull. Sci. Res. Comm. Jamaica 1 (4), 13.

Ellington, E.V., 1961b. Estimation of hypoglycin A in *Blighia sapida* (ackee). West Indian Med. J. 10, 184.

Emanuel, M.A., Benkeblia, N., 2011. Processing of ackee fruit (*Blighia sapida* L.): present and future perspectives. Acta Horticult. 894, 211–214.

Emanuel, M.A., Gutierrez-Orozco, F., Yahia, E.M., Benkeblia, N., 2013. Assessment and profiling of the fatty acids in two ackee fruit (*Blighia sapida* Köenig) varieties during different ripening stages. J. Sci. Food Agric. 93 (4), 722–726.

Evans, K.L., Arnold, L.E., 1938. Experimental studies of poisoning with ackee (*Blighia sapida*). Trans. R. Soc. Trop. Med. Hyg. 32 (3), 355–362.

Food Standards Australia New Zealand (FSANZ), New Zealand Ministry of Primary Industry and Queensland Health. Available from: <http://www.foodsafety.asn.au/resources/natural-toxins-in-food/> (accessed 18.10.14).

Foungbe, S., Naho, Y., Declume, C., 1986. Experimental study of the toxicity of arils from *Blighia sapida* (Sapindaceae) in relation to the poisoning of children of Katiola (Ivory Coast). Ann. Pharm. Fr. 44, 509–515.

Fowden, L., 1975. Hypoglycins and Related Compounds: Occurrence, Isolation, and Biosynthesis. Academic Press, New York, NY, pp. 11–19.

Fowden, L., Pratt, H.M., 1973. Cyclopropylamino acids of the genus *Acer*: distribution and biosynthesis. Phytochemistry 12, 1677–1681.

Gaillard, Y., Carlier, J., Berscht, M., Mazoyer, C., Bevalot, F., Guitton, J., et al., 2011. Fatal intoxication due to ackee (*Blighia sapida*) in Suriname and French Guyana. GC–MS detection and quantification of hypoglycin-A. Forensic Sci. Int. 206 (1), e103–e107.

Global Food Safety Initiative, 2014. Extracted on November 28, 2014. Available from: http://www.mygfsi.com/.

Golden, K.D., 2006. Hypoglycin: a toxic amino acid of the ackee plant. Caribbean Poison Information Network (CARPIN) First Scientific Conference. June 2006, Kingston, Jamaica.

Golden, K.D., Kean, E.A., 1984. The biogenesis of dicarboxylic acids in rats given hypoglycin. Biochim. Biophys. Acta 794 (1), 83–88.

Golden, K.D., Kean, E.A., Terry, S.I., 1984. Jamaican vomiting sickness: a study of two adult cases. Clin. Chim. Acta 142 (3), 293–298.

Gordon, A., 1995a. The ackee: Jamaica's enigmatic national fruit. Technical paper 950010, Grace Technology Centre, Grace, Kennedy & Co. Ltd., Kingston, Jamaica.

Gordon, A., 1995b. The truth about the ackee, Jamaica's national fruit. Presentation to Ministry of Agriculture, Fisheries and Food, United Kingdom, August 1995.

Gordon, A., 1999a. Processed ackee: a case for US market entry. Presentation to the Centre for Food Safety and Applied Nutrition. United States Food and Drug Administration, Washington, DC, April 1999.

Gordon, A., 1999b. Jamaica's ackee: a review of commercial processing and food safety. Technological Solutions Limited, Kingston, Jamaica. Technical paper prepared and submitted to the United States Food and Drug Administration, May 1999.

Gordon, A., 2011. Jamaica's ackee: challenges, opportunities and lessons for the nation. Speech to the Rotary Club of St. Andrew, Kingston, Jamaica, March 2011.

Gordon, A., 2013a. Island regulations: the impact of the Food Safety Modernization Act on Caribbean audits, imports and exports. Abstract in the Proceedings of the International Association for Food Protection (IAFP) Annual Conference, July 30, 2013, Charlotte, NC.

Gordon, A., 2013b. The Jamaican experience with the USFDA and the FSMA. Stakeholder Consultation on the Food Safety Modernization Act, exporTT, July 16, 2013, Hilton Hotel, Port of Spain, Trinidad and Tobago, WI.

Gordon, A., Jackson, J.C., 2013. The microbiological profile of Jamaican ackees (*Blighia sapida*). Nutr. Food Sci. 43 (2), 142–149.

Gordon, A., Lindsay, C., 2007. Effect of processing and handling on hypoglycin levels in ackee. CARPIN Annual Conference, Kingston, Jamaica, May 2007.

Gordon, C.L.A., 2003. Technical issues in the positioning of the Caribbean in the Free Trade Area of the Americas (FTAA). Institute of Food Technologists Annual Meeting, July 12–16, Chicago, IL, 2003 (Abstract).

Gordon, C.L.A., Jackson, J., 2002. Microbiological profile of ackee (*Blighia sapida*) during maturation. Presented at the Institute of Food Technologists Annual Conference, Anaheim, CA, USA, June 2002 (Abstract).

Hallam, D., Liu, P., Lavers, G., Pilkauskas, P., Rapsomanikis, G., Claro, J., 2004. The market for non-traditional agricultural exports. Technical paper, Raw Materials, Tropical and Horticultural Products Service Commodities and Trade Division, United Nations Food and Agricultural Organization, Rome, Italy, 2004.

Hamburg, M.A., 2014. Food Safety Modernization Act: putting the focus on prevention. Available from: www.foodsafety.gov/news/fsma.html.

Hamburg, M.A., Sharfstein, J.M., 2009. The FDA as a public health agency. N. Engl. J. Med. 360 (24), 2493–2495.

Harding, V.J., Nicholson, T.F., 1931. The nephropathic action of dicarboxylic acids on rabbits. J. Pharmacol. 42, 373–381.

Hassall, C.H., Reyle, K., 1955. Hypoglycin A and B, two biologically active polypeptides from *Blighia sapida*. Biochem. J. 60 (2), 334–339.

Hassall, C.H., Reyle, K., Feng, P., 1954. Hypoglycin A, B: biologically active polypeptides from *Blighia sapida*. Nature 173 (4399), 356–357.

Hawkes, A.D., 1972. The ackee – member of Soapberry family. The Sunday Gleaner, October 15, 1972, Kingston, Jamaica.

Henry, S.H., 1994. Toxicity of ackee fruit (*Blighia sapida*). Memo to Terry Troxell. FDA, June 28, 1994.

Henry, S.H., 2006. Update on the toxicity of ackee fruit (*Blighia sapida*). Memo to Henry Kim. FDA, March 5, 2006.

Henry, S.H., Page, S.W., Bolger, P.M., 1998. Hazard assessment of ackee fruit (*Blighia sapida*). Hum. Ecol. Risk Assess. 4, 1175–1187.

Hill, K.R., 1952. The vomiting sickness of Jamaica: a review. West Indian Med. J. 1, 243–264.

Hine, D.G., Tanaka, K., 1984a. The identification and the excretion pattern of isovaleryl glucuronide in the urine of patients with isovaleric acidemia. Pediatr. Res. 18 (6), 508–512.

Hine, D.G., Tanaka, K., 1984b. Capillary gas chromatographic/mass spectrometric analysis of abnormal metabolites in hypoglycin-treated rat urine. Biol. Mass Spectrom. 11 (7), 332–339.

Holt, C.V., Leppla, W., 1956. Study of a hypoglycemic herbal peptide. Bull. Soc. Chim. Belg. 65 (1–2), 113–123.

Hopkins, J., 1995. The glycoalkaloids: naturally of interest (but a hot potato?). Food and Chemical Toxicology 33(4), 323–328.

Howard, B.C., 2013. Amaranth: another ancient wonder food, but who will eat it? National Geographic, August 2013. Available from: http://news.nationalgeographic.com/news/2013/08/130812-amaranth-oaxaca-mexico-obesity-puente-food/ (accessed 15.11.14.).

International Food Information Council (IFIC) Foundation, 2013. 2013 Food and health survey. Available from: www.foodinsights.com (accessed 15.11.14.).

Irvine, F.A., 1930. Plants of the Gold Coast. Oxford University Press, London, 521 pp.

ITC, 2013. Market analysis and research from Trade Map. International Trade Centre (ITC). Palais des Nations, CH-1211 Geneva 10, Switzerland.

Jaffee, S.M., Henson, S., 2005. Agro-food exports from developing countries: the challenges posed by standards. Global Agricultural Trade and Developing Countries. The World Bank, Washington, DC, pp. 91–114.

Johnson, R. 2014. The U.S. trade situation for fruits and vegetable products. Congressional Research Service Report prepared for Members and Committees of Congress, RL34468, January 2014, Washington, DC.

Joskow, R., Belson, M., Vesper, H., Backer, L., Rubin, C., 2006. Ackee fruit poisoning: an outbreak investigation in Haiti 2000–2001 and review of the literature. Clin. Toxicol. (Phila) 44 (3), 267–273.

Kean, E.A., 1974. Improved method for isolation of hypoglycins A and B from fruit of *Blighia sapida*. J. Pharm. Pharmacol. 26, 639–640.

Kean, E.A., 1976. Selective inhibition of acyl-CoA dehydrogenases by a metabolite of hypoglycin. Biochim. Biophys. Acta 422 (1), 8–14.

Kean, E.A., 1988. Commentary on a review on the mechanism of ackee-induced vomiting sickness. West Indian Med. J. 37 (3), 139–142.

Kean, E.A., 1989. Hypoglycin. In: Toxicants of Plant Origin, vol. III. In: Cheeke, P.R. (Ed.), Proteins and Amino Acids. CRC Press, Boca Raton, FL.

Kean, E.A., Hare, E.R., 1980. γ-Glutamyl transpeptidase of the ackee plant. Phytochemistry 19 (2), 199–203.

Kupfer, A., Idle, J.R., 1999. Methylene blue and fatal encephalopathy from ackee fruit poisoning. Lancet 353, 1622–1624.

Lai, M.-T.D.L., Oh, E., Lui, H., 1991. Mechanistic study on the inactivation of general acyl-CoA dehydrogenase by a metabolite of hypoglycin A. J. Am. Chem. Soc. 113, 7388–7397.

Lai, M.-T.D.L., Oh, E., Lui, H-W., 1992. Studies of the inactivation of general acyl-CoA dehydrogenase by (1R)- and (1S)-(methylenecyclopropyl)acetyl-CoA. Bioorg. Med. Chem. Lett. 2, 1423–1426.

Lai, M.-T.D.L., Oh, E., Lui, H-W., 1993. Inactivation of medium-chain acyl-CoA dehydrogenase by a metabolite of hypoglycin: characterization of the major turnover product and evidence suggesting an alternative flavin modification pathway. J. Am. Chem. Soc. 115, 1619–1628.

Lancashire, R.J., Jamaican ackee. November 21, 2008. Available from: http://wwwchem.uwimona.edu.jm/lectures/ackee.html (accessed 21.07.10).

Larson, J., Vender, R., Camuto, P., 1994. Cholestatic jaundice due to ackee fruit poisoning. Am. J. Gastroenterol. 89 (9), 1577–1578.

Lewis, C.B., 1961. The ackee – a brief historical note. Bull. Sci. Res. Comm. Jamaica 1 (4), 12.

Manchester, J.E., Manchester, K.L., 1980. Separation of hypoglycin A from leucine and other amino acids on Sephadex G-10. J. Chromatogr. 193, 148.

Manchester, K.L., 1974. Biochemistry of hypoglycin. FEBS Lett. 40, 133.

Martin-Wilkins, A. 2005. 400M Jamaican ackee market under threat. The Jamaica Observer, December 29, 2005, Kingston, Jamaica.

McGowan, C., Wiley, V., Bates, R.P., 1989a. Application of methodology for RP-HPLC amino acid analysis to the measurement of hypoglycin A. Int. J. Bio-Chromatogr. 4 (3), 161.

McGowan, C., Wiley, V., Bates, R.P., 1989b. Use of OPT-HPLC amino acid analysis for the quantitation of the toxin, hypoglycin B, and other amino acids in the Jamaican ackee fruit. Fed. Proc. 3 (4), 5963.

McTague, J.A., Forney, Jr., R., 1994. Jamaican vomiting sickness in Toledo. Ohio. Ann. Emerg. Med. 23 (6), 1116–1118.

Meda, H.A., Diallo, B., Buchet, J.P., Lison, D., Barennes, H., Ouangré, A., et al., 1999. Epidemic of fatal encephalopathy in preschool children in Burkina Faso and consumption of unripe ackee (*Blighia sapida*) fruit. Lancet 353 (9152), 536–540.

Mendonca, A., Gordon, C.L.A., 2004. Control of food safety and quality in Caribbean countries: implications for international trade. Institute of Food Technologists Annual Meeting, July 12–16, Las Vegas, NV, 2004 (Abstract).

Morton, J.F., 1987. Fruits of Warm Climates. Florida Flair Books, Miami, FL, USA.

Moya, J., 2006. Ackee (*Blighia sapida*) poisoning in the Northern Province, Haiti, 2001. Epidemiol. Bull. 22 (2), 8–9.

National Geographic, 2011. Top 10 national dishes. Extracted on October 15, 2014. Available from: http://travel.nationalgeographic.com/travel/top-10/national-food-dishes/.

Odutuga, A.A., Asemota, H.N., Musac, I., Gloden, K.D., Kean, E.A., 1992. Fatty acid composition of arilli from ackee fruit (*Blighia sapida* L.). Jamaican J. Sci. Technol. 3, 30–32.

Patrick, S.J., Jelliffe, D.B., Stuart, K.L., 1955. The hepatic glycogen content in acute toxic hypoglycemia. J. Trop. Pediatr. 1 (2), 88.

Perkins, K.D., Payne, W.W., 1978. Guide to Poisonous and Irritant Plants of Florida. University of Florida and Florida State Museum, Gainesville, FL.

Pflug, I.J., 1987. Factors important in determining the heat process value, FT, for low-acid canned foods. J. Food Prot. 50 (6), 528–533.

Potter, N.N., Hotchkiss, J.H., 1995. Vegetables and fruits. In: Potter, N., Hotchkiss, J. (Eds.), Food Science. fifth ed. Aspen Publishers Inc., Maryland, USA. pp. 409–436.

Regmi, A., 2001. Changing structure of global food consumption and trade. In: Regmi, A. (Ed.), Agriculture and Trade Report WRS-01-1, Market and Trade Economics Division. Economic Research Service, U.S. Department of Agriculture, Washington, DC.

Rizzoli, A.A., Galzigna, L., 1970. Molecular mechanism of unconscious state induced by butyrate. Biochem. Pharmacol. 19 (10), 2727–2736.

Saltsman, J., DeVlieger, D., 2012. FDA's Outreach Program with Jamaican Authorities and Ackee Processors. Internal FDA Report on the Jamaica Ackee Outreach Project. United States Food and Drug Administration, Silver Spring, Maryland, USA, July 2, 2012.

Sarwar, G., Botting, H.G., 1994. Reversed-phase liquid chromatography determination of hypoglycin A (HG-A) in canned ackee fruit samples. J. AOAC Int. 77, 1175–1179.

Sarwar, G., Botting, H.G., Peace, R.W., 1988. Complete amino acid analysis in hydrolysates of foods and feeds by liquid chromatography of precolumn phenylisothiocyanate derivatives. J. Assoc. Off. Anal. Chem. 71, 1172–1175.

Scott, H.H., 1917. The vomiting sickness of Jamaica. Trans. R. Soc. Trop. Med. Hyg. 10 (3), 47–66.

Scott, P.M., Botting, H.G., Kennedy, B.P.C., Knipfel, J., 1974. The determination of hypoglycin A in ackee. J. Food Sci. 39 (5), 1057–1058.

Seaforth, C.E., 1962. The ackee – Jamaica's national fruit. Bull. Sci. Res. Counc. 3 (1–4), 51–53.

Senior, O., 1983. A–Z of Jamaican Heritage. Heinemann Educational Books (Caribbean) and Gleaner Co., Kingston, Jamaica.

Senior, A.E., Sherratt, H.S.A., 1968. Biochemical effects of hypoglycemic compound pent-4-enoic acid and related non-hypoglycemic fatty acids: carbohydrate metabolism. Biochem. J. 110, 521–527.

Sherratt, H.S.A., 1986. Hypoglycin, the famous toxin of the unripe Jamaican ackee fruit. Trends Pharmacol. Sci. 7, 186–191.

Sherratt, H.S.A., 1969. Hypoglycin and related hypoglycaemic compounds. Br. Med. Bull. 25, 250–255.

Shih, V., Tanaka, K., 1978. Plasma and urine amino acid changes in rats treated with hypoglycin. Clin. Chim. Acta 88 (3), 539.

Singh, P., Gardner, M., Poddar, S., 1992. Toxic effects of ackee oil (*Blighia sapida* L.) following subacute administration to rats. West Indian Med. J. 41 (1), 23–26.

Sloan, E., 2014. Top ten functional food trends. Food Technol. 68 (4).

SRC, 1999. Ackee – Development of an Export Trade. SRC Press, Scientific Research Council, Kingston, Jamaica.

STATIN, 2014. International merchandise trade statistical bulletin, January to December 2013. Extracted on October 20, 2014. Available from: http://statinja.gov.jm/Trade-Econ%20Statistics/InternationalMerchandiseTrade/newtrade.aspx.

Stuart, K.L., 1975. Vomiting sickness of Jamaica. In: Pan American Association of Biochemical Societies (PAABS) Symposium Series, vol. 3. In: Kean, E.A. (Ed.), Hypoglycin. Academic Press, New York, pp. 39–44.

Stumbo, C.R., 1973. Thermobacteriology in Food Processing. Academic Press, Inc, San Diego, CA.

The Daily Gleaner, 2000. US lifts ackee ban. The Daily Gleaner, Friday, July 7, 2000, Kingston Jamaica.

Tanaka, K., 1972. On the mode of action of hypoglycin A. J. Biol. Chem. 247, 7465–7478.

Tanaka, K., 1975. Branched pentanoic acidemia and medium chain dicarboxylic aciduria induced by hypoglycin A. In: Kean, E.A. (Ed.), Hypoglycin. Academic Press, New York, p. 67.

Tanaka, K., 1979. Jamaica vomiting sickness. In: Vinkin, P.J., Bryan, G.W. (Eds.), Handbook of Clinical Neurology, vol. 37 (Part II). North Holland, Amsterdam.

Tanaka, K., Ikeda, Y., 1990. Hypoglycin and Jamaica vomiting sickness. In: Fatty Acids Oxidation: Clinical, Biochemical and Molecular Aspects. Alan R. Liss, Inc., New York, p. 167.

Tanaka, K., Isselbacher, K.J., Shih, V., 1972. Isovaleric and alpha-methylbutyric acidosis induced by hypoglycin A: mechanism of Jamaican vomiting sickness. Science 175, 69–71.

Tanaka, K., Kean, E.A., Johnson, B., 1976. Jamaican vomiting sickness: biochemical investigation of two cases. N. Engl. J. Med. 295 (9), 461–467.

Tanaka, K., Miller, E.M., Isselbacher, K.J., 1971. Hypoglycin A: a specific inhibitor of isovaleryl CoA dehydrogenase. Proc. Natl Acad. Sci. USA 68 (1), 20–24.

United States Department of Agriculture (USDA), 2014. Importation of kiwi from Chile into the United States, a proposed rule. Available from: https://www.federalregister.gov/agencies/animal-and-plant-health-inspection-service;AnimalandPlantHealthInspectionService,USDA.Available from: https://www.federalregister.gov/articles/2014/10/16 (accessed 16.10.14.). Extracted from http://www.ers.usda.gov/topics/crops/fruit-tree-nuts/trade.aspx on January 28, 2015.

United States Department of Agriculture (USDA), 2015. USDA National Nutrient Database for Standard Reference, Release 27. Extracted on January 28, 2015. Available from: http://ndb.nal.usda.gov/ndb/search/list.

United States (US) Food and Drug Administration (FDA), 2000. FDA Import alert IA #21-11, 7/3/00, Available from: http://www.fda.gov/ora/fiars/ora_import_ia2111.html (accessed 21.07.10.).

US FDA, 2010a. Haitian ackee fruit. January 2010. Available from: http://www.fda.gov/Food/NewsEvents/WhatsNewinFood/ucm197850.htm (accessed 21.07.10).

US FDA, 2010b. Detention without physical examination of ackees. Import Alert 21–11. June 3, 2010. Available from: http://www.accessdata.fda.gov/cms_ia/importalert_64.html (accessed 21.07.10).

US FDA, 2014a. The FDA FSMA – key facts. Extracted on June 23, 2014. Available from: http://www.fda.gov/Food/GuidanceRegulation/FSMA/ucm237934.htm.

US FDA, 2014b. FSMA proposed rule for preventive controls for human food. Extracted on June 23, 2014. Available from: http://www.fda.gov/downloads/Food/GuidanceRegulation/FSMA/UCM360735.pdf.

US FDA, 2014c. Proposed final rule – produce. Extracted on June 23, 2014. Available from: http://www.fda.gov/downloads/food/guidanceregulation/fsma/ucm360734.pdf.

US FDA, 2014d. Overview of the FSMA proposed rules on produce safety and standards and preventive controls for human food. Extracted on June 23, 2014. Available from: http://www.fda.gov/Food/GuidanceRegulation/FSMA/ucm334120.htm.

US FDA, 2014e. FSMA proposed rule – Foreign Supplier Verification Program. Extracted on June 23, 2014. Available from: http://www.fda.gov/food/guidanceregulation/fsma/ucm361902.htm.

US FDA, 2014f. FDA form 843 frequently asked questions. Available from: http://www.fda.gov/iceci/inspections/ucm256377.htm (accessed 23.06.14).

Vargas, E.A., Whitaker, T.B., Santos, E.A., Slate, A.B., Lima, F.B., Franca, C.A., 2006. Design of sampling plans to detect ochratoxin A in green coffee. J. Food Addit. Contam. 23, 62–72.

von Holt, C., 1966. Methylenecyclopropaneacetic acid, a metabolite of hypoglycin. Biochim. Biophys. Acta 125 (1), 1–10.

von Holt, C., von Holt, M., Bohm, H., 1966. Metabolic effects of hypoglycin and methylenecyclopropylacetic acid. Biochim. Biophys. Acta 125 (1), 11–21.

Walters-Gregory, S. 2012. New food standard for Jamaica. The Daily Gleaner, October 5, 2012, Kingston, Jamaica.

Ware, G.M., 2002. Method validation study of hypoglycin A determination in ackee fruit. J. Assoc. Off. Anal. Chem. Int. 85, 933–937.

Ware, G., 2008. Summary report: Jamaica ackee laboratories evaluation. Report of a FDA study to evaluate the analytical capability of laboratories responsible for determining the levels of hypoglycin A in exported ackee. United States Food and Drug Administration, Bethesda, MD, USA, August, 2008.

WebMD, 2015. Graviola: uses, side-effects, interaction, dosing. Extracted on November 15, 2014. Available from: http://www.webmd.com/vitamins-supplements/ingredientmono-1054-GRAVIOLA.aspx?activeIngredientId=1054&activeIngredientName=GRAVIOLA.

Whitaker, T.B., Saltsman, J.J., Ware, G.M., Slate, A.B., 2007. Evaluating the performance of sampling plans to detect hypoglycin A in ackee fruit shipments imported into the United States. J. AOAC Int. 90 (4), 1060–1072.

Wilson, N., 2011. Ackee kills 23, sickens 194 in less than 3 months. The Jamaica Observer, February 24, 2011, Kingston, Jamaica.

SUBJECT INDEX

H

I

Printed in the United States
By Bookmasters